THE STORY
OF
THE EARTH

THE STORY OF THE EARTH

OF

THE EARTH

Peter Cattermole and Patrick Moore

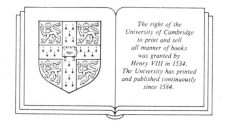

The right of the
University of Cambridge
to print and sell
all manner of books
was granted by
Henry VIII in 1534.
The University has printed
and published continuously
since 1584.

Cambridge University Press
Cambridge
London   New York   New Rochelle
Melbourne   Sydney

This book was designed and produced by
The Oregon Press Limited, Faraday House.
8 Charing Cross Road, London WC2H 0HG

Published by the Press Syndicate of the University
of Cambridge, The Pitt Building, Trumpington Street,
Cambridge CB2 1RP
32 East 57th Street, New York, NY 10022, USA
296 Beaconsfield Parade, Middle Park, Melbourne 3206,
Australia

First published in 1985

Filmset by SX Composing Limited, Rayleigh, England
Printed and bound by Printer Industrai Grafica SA, Barcelona

Library of Congress catalogue card number: 84-17072

British Library Cataloguing in Publication Data

Cattermole, Peter
The story of the earth.
1. Historical geology
I. Title   II. Moore, Patrick
551   QE28.3

ISBN 0 521 26292 5
D.L.B. 33494-1984
Artwork: Paul Doherty

OPPOSITE TITLE-PAGE
Eocene strata etched into 'badlands', South Dakota.

Contents

Part One

BEGINNINGS

Sandstone spires, Bryce
Canyon National Park, Utah.

1

PLANET EARTH

The Earth is our home in space. Ancient civilizations believed it to be flat, with the heavens revolving round it once a day; even when the Greeks showed that the Earth was a globe, it was still thought to be the most important body in the universe. This was natural enough, but the idea of the Earth's supreme importance persisted for a surprisingly long time. As recently as 1600 the Italian philosopher Giordano Bruno was burned at the stake in Rome, one of his heretical beliefs being that the Earth moves round the Sun. The final proof that the Earth is an ordinary planet did not come until the seventeenth century.

Studies of the Earth involve many branches of science. The geologist has to call upon the physicist, the chemist, the meteorologist and the astronomer. Then, too, there is the problem of how life began, and there are many conflicting views (quite apart from the biblical fundamentalists). It is generally believed that life on Earth began on the Earth itself, some time after the formation of the planets between four and five thousand million years ago, but it has also been suggested that life was deposited here from outer space. Whether or not this is true, we have no definite evidence of life elsewhere. The other planets in the Sun's family are unsuitable in their various ways, but the Sun is only one of a hundred thousand million stars in our Galaxy – and the Galaxy itself is one of many. It seems both conceited and illogical to assume that mankind is unique.

We have solved many of the problems that seemed baffling not so very long ago. We have a good idea of how the Earth was formed, and we can fix its age with reasonable accuracy; the record of the rocks can tell us much about its past history, and we can use the methods of palaeontology, or fossil studies, to tell us what the world used to be like hundreds of millions of years ago. We have mapped the whole of the Earth's surface; we have explored the sea floor, and we have sent our instruments into and above the atmosphere. Of course, even modern science has marked limitations, and this was shown forcibly in the early 1970s, when an attempt was made to bore through the Earth's crust and obtain samples from the layers below. 'Project Mohole' was an expensive failure, and we still have no direct knowledge of conditions at the Earth's core, even though theory can give us what seems to be at least a reasonably satisfactory picture.

At least we can now do what would have seemed inconceivable at the start of the century: look at the Earth from space, and from the surface of the Moon. It is these views, perhaps more than anything else, that have made us realize that the Earth is a planet, unexceptional in the solar system except that it alone is suited to our kind of life.

The Earth in the Solar System

A casual glance at a plan of the solar system shows that it is divided into two parts. First come four relatively small, solid planets: Mercury,

Venus, the Earth and Mars. Beyond the orbit of Mars there is a wide gap, in which move thousands of dwarf worlds known as minor planets or asteroids. These are followed by the four giants: Jupiter, Saturn, Uranus and Neptune, which have gaseous surfaces and are quite unlike the Earth. Finally there is Pluto, a curious little world which is now known to be smaller and less massive than the Moon, so that it may not be worthy of true planetary status. Their revolution periods range from 88 days for Mercury up to 248 years for Pluto: although when near its perihelion (the point at which its path is closest to the Sun) Pluto's eccentric orbit brings it within that of its giant neighbour, Neptune.

Of the inner planets, Mercury is the smallest, with a diameter of 4840 km. It has a cratered surface, superficially very like that of the Moon, and it has virtually no atmosphere. Venus is very nearly as large as the Earth; it has a diameter of 12,200 km, and has an almost circular orbit at a distance of 108,000 km from the Sun. Yet Venus and the Earth are not in the least similar. Venus has a dense atmosphere made up chiefly of carbon dioxide; the ground pressure is at least 90 times that of the Earth's air at sea level, and the temperature is 500°C. The clouds contain large amounts of sulphuric acid, and in every way Venus appears to be overwhelmingly hostile. Mars, at 227,000 km from the Sun, has a diameter of 6760 km and a revolution period of 687 days. Although much smaller than the Earth it is less hostile than Venus, and has a thin, carbon dioxide atmosphere. The Martian poles are covered by white caps, made up of a mixture of water ice and carbon dioxide ice. The inclination of Mars' axis is almost the same as that of the Earth, and so the seasons are of the same type, apart from being much longer. However, the atmospheric pressure is too low for seas to exist, and present evidence indicates that there is no life there. The surface is cratered, and there are high volcanoes, one of which (Olympus Mons) is three times the height of Everest.

The zone around the Sun in which temperatures are suitable for the existence of life, provided that other conditions are fulfilled, is often termed the ecosphere. Venus lies at the inner edge of the ecosphere and Mars at the outer edge, while the Earth moves in the centre of the zone. It is likely that in the early days of the solar system, when the Sun was 30 per cent less luminous than it is now, Venus and the Earth started to evolve along similar lines; but when the Sun's luminosity increased, the oceans of Venus evaporated and the carbonates were driven out of the rocks, turning the planet into the furnace-like environment of today.

Although the giant planets are so massive (the mass of Jupiter is over 300 times that of the Earth) they can have no effects upon conditions on our world simply because they are so far away. However, there are some asteroids that move away from the main swarm and approach the Earth; in 1937 a tiny asteroid, Hermes, passed by at less than twice the distance of the Moon. Occasional collisions are to be expected, and have no doubt occurred in the past. It has even been suggested that an asteroid impact 65,000,000 years ago led to a climatic change so violent that it caused the extinction of many forms of life, including the dinosaurs. However, positive proof is lacking, and the chances of a major impact in the foreseeable future are slight.

Yet the Earth will not last for ever. Eventually the Sun will change its structure, and for a time will send out 100 times as much radiation as it does now. This will certainly mean the end of life on Earth, if not of the

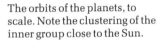

The orbits of the planets, to scale. Note the clustering of the inner group close to the Sun.

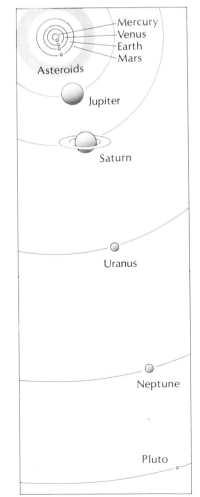

Earth itself; but as the change in the Sun will not occur for at least 4000 million years, and probably longer, there is no immediate cause for alarm on this score!

The Movements of the Earth

The Earth is the third planet of the solar system. In round figures, its distance from the Sun is 150,000,000 km, but its path or orbit is not circular; the distance ranges between 147,000,000 km at closest approach (perihelion) out to 153,000,000 km at furthest recession (aphelion). The revolution period is, of course, one year – more precisely 365.25 days; it is the extra quarter of a day that is responsible for the complications in our calendar. To prevent the civil calendar from becoming 'out of step' with the seasons, an extra day is added every four years; in these Leap Years February has 29 days instead of its usual 28. A further refinement in our present calendar is to omit 'century years' as Leap Years unless exactly divisible by 400. Thus 1900 was not a Leap Year, but 2000 will be.

The seasons have little connection with the changing distance from the Sun. Perihelion actually occurs in December, when it is winter in the northern hemisphere. The Earth's axis is inclined to the perpendicular to the orbital plane by an angle of 23.5 degrees; in northern summer the North Pole is tilted toward the Sun, while in northern winter it is the turn of the South Pole to receive the maximum benefit of the Sun's rays. Theoretically, the southern winters should be longer and colder than the northern, and the summers should be shorter and hotter, since the Earth moves fastest when near perihelion; but the difference is more or less cancelled out by the greater amount of continent-free ocean in the southern hemisphere, which tends to stabilize the temperature.

The axial inclination is not constant. Over long periods it varies perceptibly, because of the combined effects of the Sun and the Moon; the Earth is not a perfect sphere, and the gravitational pulls upon the equatorial 'bulge' are responsible for the changes in inclination. Over the past 400 years the change has amounted to about half a degree. The direction of the spin axis also changes slowly but regularly, describing a full circle in a period of 26,000 years. This is termed precession. At present the axis points northward, to a position in the sky marked within a degree by a bright star known as Polaris or the Pole Star, in the constellation of Ursa Minor, the Little Bear; but when the Egyptian Pyramids were built, the north celestial pole lay close to a much fainter star, Thuban in Draco (the Dragon). In 12,000 years' time the north pole star will be the brilliant blue Vega. There is no conspicuous south polar star – a fact which past ocean navigators in the southern hemisphere have had cause to regret.

Neither is the length of the day absolutely constant. Tidal friction between the Earth and the Moon causes a slight but definite lengthening, easily measurable with modern atomic clocks – which are better timekeepers than the Earth itself. The average increase in length amounts to a tiny fraction of a second per century, but in past geological periods the rotation period was considerably longer than it is now. There are also small random variations, due presumably to movements deep inside the Earth's globe.

The orbital eccentricity also varies slightly, but the Earth's path is stable enough; there is no overall approach to or recession from the Sun. This is fortunate for us, since even a slight variation would have

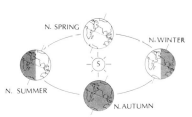

The seasons result from the planet's axial inclination of 23.5 degrees to the plane of the orbit. When the northern hemisphere tilts toward the Sun it has its summer while the southern hemisphere experiences winter.

marked effects upon our climate. Were the Earth even ten million kilometres closer to the Sun, the whole story of evolution would have been very different.

The Earth's Atmosphere

In the solar system, the Earth is unique in one respect: it is the only planet to have an atmosphere that we could breathe. Most of it is made up of nitrogen (78.08 per cent) and oxygen (20.95 per cent). Next comes argon (0.93 per cent), with much smaller amounts of other gases such as carbon dioxide, neon, helium and hydrogen.

However, the present atmosphere is not the original one. In the early period of the Earth's existence there was probably a hydrogen-rich atmosphere which was stripped away by the influence of the Sun, and the atmosphere we know today was produced from inside the Earth itself. At first it contained much more carbon dioxide and much less free oxygen than it does now. It was only when plant life developed, perhaps 2000 million years ago, that carbon dioxide was removed and oxygen produced; plants take in carbon dioxide and give out oxygen by the well-known process of photosynthesis.

The bottom part of the atmosphere is known as the troposphere: it extends upward for an average height of about 12 km, and it is here that we find all our normal clouds and weather. The temperature decreases with height, as every mountaineer and airman knows. Above the troposphere comes the tropopause, which is in turn succeeded by the stratosphere, extending up to about 80 km. The temperature is more or less constant, and here we find the layer of ozone (O_3), a special form of oxygen which acts as a shield against harmful shortwave radiations from space; without the ozone layer it is unlikely that life on Earth could ever have developed. The upper part of the stratosphere is often termed the mesosphere.

Next comes the ionosphere, containing the layers that reflect some radio waves and make long-range radio communication possible. Finally there is the exosphere, which has no definite boundary, but simply thins out until its density is no greater than that of the interplanetary medium.

Part of the ionosphere is sometimes called the thermosphere and is characterized by a very high temperature. This does not, however, mean that it is hot in the ordinary meaning of the term. Temperature is determined by the velocities at which the atoms and molecules move around; the faster the movements, the higher the temperature. In the thermosphere the movements are very rapid, and so the temperature is high, but there are so few atoms and molecules that the actual quantity of heat is inappreciable. There is a good analogy with a glowing poker and a firework sparkler. Each spark is white-hot, but contains so little mass that there is no danger in holding the firework by hand – but it would be distinctly unwise to pick up a red-hot poker, whose actual temperature is much lower.

Aurorae or polar lights are produced in the upper atmosphere. Solar wind particles enter the Van Allen zones and 'overload' them, so that charged particles cascade down toward the magnetic poles and produce the lovely auroral displays. Aurorae are most frequent and most brilliant when the Sun is at its most active, which happens approximately every 11 years (the last solar maximum was in 1980, so that the next may be expected around 1991). Meteors are tiny particles which dash into the upper air and burn away by friction; generally

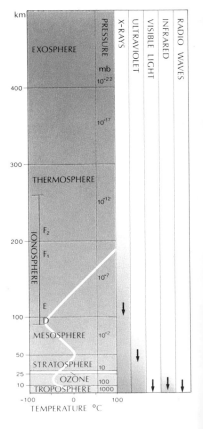

Structure of the atmosphere. Variations in atmospheric temperature and pressure result from the distribution of solar heating. The troposphere is the zone of our weather and is separated from the stratosphere by the tropopause. It is within the stratosphere and the higher rarified layers that most of the ultraviolet radiation is absorbed.

they become luminous at about 190 km above the surface and burn out at 65 km or above. The atmosphere is no effective screen against larger masses, known as meteorites, which reach the ground and may cause craters such as the famous Meteor Crater in Arizona; but missiles of this size are rare, and there is no reliable record of any human death caused by a meteorite fall.

The Earth and the Moon

The differences between Earth and Moon are largely a function of the Moon's much lower mass. The escape velocity is too low for the retention of an atmosphere, and neither are there any traces of hydrated materials: that is to say, materials that have contained H_2O in any form. The Moon is a lifeless world, and has always been so. The so-called seas are of lava, not water or ice; there are high peaks, together with ridges and valleys, while the whole lunar scene is dominated by the craters, some of which are more than 300 km in diameter. By geological standards, the lunar formations are very old. Little has happened on the Moon's surface over the past 3000 million years or so.

The Moon has a diameter of 3475 km and a mass 1/81 that of the Earth. It has a mean distance from the Earth of 384,000 km and its revolution period is 27.3 days, although because the Earth-Moon system is in orbit around the Sun the lunar synodic period (that is to say, the interval between one new moon and the next) is 29.5 days. The phases depend upon the amount of the sunlit hemisphere that is turned toward us. When the Moon is between the Sun and the Earth, it is 'new', and cannot be seen – except when the alignment is perfect, when the Moon passes in front of the Sun and produces a solar eclipse. This does not happen at every new moon, because the Moon's orbit is appreciably inclined to that of the Earth. If, when it is full, the Moon passes into the Earth's shadow, we see a lunar eclipse.

To be strictly accurate, the Moon and the Earth move round their common centre of gravity, or barycentre. However, the great difference in mass means that the barycentre lies deep inside the Earth's globe, so that the simple statement that 'the Moon goes round the Earth' is good enough for most purposes.

The Moon is the controlling force producing the tides. The basic principle is straightforward enough. The Moon pulls upon the Earth and tends to produce a 'bulge'; water is more easily affected than solid ground, so that water heaps up under the Moon at a high tide, with a similar high tide on the far side of the Earth. As the Earth rotates, the 'bulges' do not rotate with it, and so the high tides sweep round the globe. In practice the whole situation is extremely complicated, and we must also reckon with the tide-producing force of the Sun, which is weaker than that of the Moon but is nevertheless very evident. When the Sun and Moon are pulling in the same sense (at new or full moon) we have strong or 'spring' tides; when the two are pulling against each other (at half-moon) we have weaker or 'neap' tides.

It is tidal friction that is slowly increasing the length of the day. In past ages, before the Moon solidified, tides in it produced by the Earth slowed the rotation so much that, relative to the Earth, it stopped rotating altogether; today the Moon keeps the same face turned to the Earth all the time. It should be borne in mind, however, that the Moon does not keep the same face turned toward the Sun, so that day and night conditions are the same all over the lunar surface.

2

THE FORMATION
OF THE EARTH

Early Notions

What can we really say about the Creation? Probably no subject has caused more discussion – and more dissent – over the years, and even today we cannot claim to know much about the beginning of the universe itself. But so far as the Earth is concerned, we have at least a few firm facts to guide us, and in particular we are reasonably confident about the time scale. The Earth began its separate existence about 4700 million years ago; the universe is older still, and a reasonable estimate leads us to a figure of around 15,000 million years. According to the current school of thought, all matter came into existence suddenly, in what is termed the Big Bang. Once created, the matter began to expand; in the course of time galaxies were formed, then stars, and then planets.

What we do not know is the reason for the initial Big Bang. The material must have come from somewhere. If it originated in a Big Bang, then what happened before that? The only alternative is to assume that the universe has always existed, and will never die. In the late 1940s the theory of continuous creation was put forward by a group of scientists at Cambridge University, and was later refined and elaborated by Sir Fred Hoyle; the expansion of the universe was accepted, but it was claimed that old galaxies, passing beyond observable range, would be replaced by new ones, created from material which appeared spontaneously in the form of hydrogen atoms. This so-called Steady State theory was popular for a time, but gradually the evidence against it mounted up, and today it has been almost totally abandoned. The only logical alternative to the Big Bang is the oscillating universe, assuming that the present phase of expansion will be followed by one of contraction until all the material comes together again, producing a new Big Bang.

Going back to the seventeenth century, we find that James Ussher, Anglican Archbishop of Armagh, solved the whole problem neatly by adding up the ages of the Old Testament patriarchs and making similar calculations which satisfied him and many others. In 1650 he stated categorically that the world had begun its career at ten o'clock in the morning of 26 October, 4004 BC. Subsequently it was shown beyond doubt that many fossils are much older than that, and a more scientific explanation was sought.

Early Theories

One interesting attempt at explaining the Earth's origin was made by a Frenchman, the Comte de Buffon, in 1779. He argued that the Earth must be at least 75,000 years old, and could have been produced by a collision between the Sun and a comet, with the result that the material was sprayed around in sufficient quantities to form planets. While this idea is very wide of the mark, since comets are very flimsy objects that could not possibly cause such a catastrophe, at least he improved the time scale a little. Furthermore, his contemporary, Mikhail Lomonosov, Russia's first great astronomer, was considering an age of hundreds of thousands of years. Different ideas were put forward by the philosophers Immanuel Kant in Germany and Thomas Wright in England, who believed that the planets were formed from a cloud of material associated with a youthful Sun. This type of 'Nebular Hypothesis' was refined and improved in 1796 by Pierre Simon, Marquis de Laplace, who was an excellent mathematician as well as a far-sighted theorist.

According to Laplace, the solar system began as a vast cloud of gas, disk-shaped and in slow rotation. Gravity caused it to contract; as it did so, its rate of rotation increased, until the centrifugal force at its edge became equal to the gravitational pull there. At this stage a ring of matter broke away from the main mass, and collected into a planet. As the contraction continued, a second ring was thrown off, and so on. The planets were formed one by one, and the remnant of the original cloud made up the present-day Sun.

It all looked very neat and tidy, and Laplace's Nebular Hypothesis was not challenged for many years, but gradually some awkward facts emerged. Mathematical analyses showed that rings would not be thrown off during contraction of the cloud; and even if this were possible, these rings could not condense into planets. Worse still was the problem of angular momentum, a quantity that is obtained by considering one body moving around another and combining its mass, distance and velocity. It is a fundamental principle that while angular momentum can be transferred, it can never be destroyed. On Laplace's theory all the angular momentum possessed by the Sun and planets must originally have been contained in the gas cloud, and most of it would now be concentrated in the Sun. This meant that the Sun should be rotating fairly quickly. Actually, the Sun's rotation period is 25 days, and almost all of the angular momentum of the solar system resides in the four giant planets Jupiter, Saturn, Uranus and Neptune. There seemed to be no way of overcoming this difficulty, and the Nebular Hypothesis was reluctantly discarded, to be replaced by a crop of 'catastrophic' theories involving the action of a passing star.

The first of these was propounded by two Americans, the geologist Thomas Chamberlin and the astronomer Forest Ray Moulton, in 1900. Space is sparsely populated. The Sun, which is a normal star, has a diameter of 1,392,000 km. However, if we were to represent it by a tennis ball its nearest stellar neighbour would reside over 1000 km away. Therefore, collisions or even close encounters of two stars must be very rare, at least in our part of the universe. However, we cannot entirely rule out an encounter, and the two Americans believed that as a wandering star approached the Sun, a cigar-shaped tongue of material was torn away from the solar surface. As the intruder receded, the tongue of matter was left whirling round the Sun; slowly it condensed into planets, with the largest bodies, Jupiter and Saturn, in the middle of the solar system, where the thickest part of the 'cigar' would have been.

The passing-star theory was supported and popularized by Sir James Jeans, remembered today not only for his contributions to theoretical astrophysics but also for his books and broadcasts. It, too, looked

Diagram to show the condensation of the planets from a rapidly rotating solar cloud.

remarkably neat; it would mean that planetary systems would be very common in the Galaxy. Yet again there were mathematical problems, and attempts at further refinement were no more successful. The best of them came from Sir Harold Jeffreys, who suggested that the passing star struck the Sun a glancing blow.

The next theories involved a binary companion to the Sun. Star pairs or binaries are very common, and there was nothing outrageous in suggesting that the Sun also might once have been one component of a binary system. The American astronomer Henry Norris Russell believed that it was the companion star that was struck by the intruder, causing enough debris to form the planets; R. A. Lyttelton considered that a near approach from the wanderer would wrench the binary companion away from the Sun, planet-forming material being scattered in the process; Sir Fred Hoyle proposed that the solar companion exploded as a supernova, shedding much of its material in the Sun's neighbourhood before departing permanently.

An interesting modification was due to G. P. Kuiper, who regarded the solar system as a 'degenerate binary', in which the second mass did not condense into a star, but remained as material scattered around. Therefore, the end product was one star (the Sun) and several condensations or 'protoplanets', which gradually formed into the planets of today. Once a protoplanet became sufficiently massive, it would draw in material by its own gravitational pull.

Modern Hypotheses

Recent ideas are different. Passing stars and binary companions are generally out of favour, but mention should be made of the theory proposed by Michael Woolfson of York University, who suggests that the Sun was produced as a member of a whole cluster of stars, relatively close together. If the Sun's formation were completed at a fairly early stage in the history of the cluster, it could have been approached by a 'protostar' at an earlier stage in its evolution; the Sun then pulled a long filament of material away from the protostar, and this filament broke up into drops from which the planets were formed. Woolfson's hypothesis is by no means unreasonable, but in general most modern theories are much more like Laplace's Nebular Hypothesis than those of the catastrophists.

The trend was set in the 1950s by Otto Schmidt in Russia and Carl von Weizsäcker in Germany, whose ideas were similar in principle though different in detail. This time it was assumed that the young Sun was surrounded by a 'solar nebula', a huge envelope of material made up largely of the two lightest elements, hydrogen and helium, together with what may be termed dust. Collision and friction between the Sun's family of particles led to the formation of a disk-shaped cloud, from which the planets were built up by the less than perfectly understood process of accretion.

Basically, this is still supported today, and it is usually believed that the solar system did indeed begin as a contracting solar nebula. Rotation meant that the cloud would become disk-shaped and that planets would then form by accretion, although not in the way that Laplace had supposed. The outermost part of the cloud was naturally the coldest, and substances such as water, ammonia, methane and so on solidified; the embryo outer planets consisted mainly of these materials, and when they became sufficiently massive they could pull in the gaseous hydrogen and helium, ending up as the four giant planets. Closer to the Sun, things were different. Here, the temperature was higher, and planetary growth followed a different pattern. While the temperatures prevailing in the solar nebula remained high (around 1000°C), dense, iron-rich compounds precipitated in the proto-inner planets to form the dense planetary cores. When temperatures had dropped sufficiently, silicate materials of lower density could be swept up by the growing protoplanets. This explains why the solar system is divided so definitely into two parts. The fact that there is no major planet between the orbits of Mars and Jupiter can be accounted for by the powerful pull of Jupiter, which disrupted any embryo planet in this region before it had a chance to grow, producing instead the swarm of dwarf worlds which we call asteroids.

If this is correct, we may expect all the planets to be of roughly the same age, and it is certainly true of the Moon, which is probably better regarded as a companion planet rather than as a mere satellite of the Earth. Not everyone agrees with the theory. A. J. R. Prentice of Monash University in the United States has reverted to a modified form of the old Nebular Hypothesis, which he believes is a real possibility, provided that the solar nebula collapsed quickly enough: in which case the giant planets would be older than the Earth. The Swedish physicist Hannes Alfvén has emphasized the importance of magnetic fields, and believes that in its early days the Sun was fast-spinning but without planets; later, its strong magnetic field resulted in the sucking-in of material, which condensed into planets and at the same time slowed down the Sun's rotation.

Quite clearly there are still marked divergencies of opinion, but one thing does seem more or less established: the Earth and the other planets were born between 4500 and 5000 million years ago from a solar nebula, and that the smaller inner planets were hot from a very early stage. It also seems certain that in the early stages of the solar system the Sun was less luminous than it is now, perhaps by a factor of at least 25 per cent. It was also somewhat unstable, and ejected material in the form of a strong 'solar wind'. This, too, is supported by observational evidence, since we know of very young stars, known as T Tauri stars, which are in this stage now. The T Tauri wind would have blown much of the thinly spread material away from the inner part of the solar system, and would also take most of the total angular momentum away from the Sun.

There is one very interesting result of this general picture. Venus and the Earth are similar in size and mass, and it has been suggested that they began to evolve along similar lines, developing solid surfaces, then oceans, and then similar types of atmosphere. But as the Sun gradually became more luminous, the situation changed. The Earth was sufficiently far out to avoid the worst effects of the increased luminosity; Venus was not – and so the oceans of Venus boiled and evaporated, the temperature became intolerably high, and carbonates were driven out of the rocks, producing the thick, carbon dioxide atmosphere that Venus has today. Carbon dioxide acts in the manner of a greenhouse, shutting in the Sun's heat, so that on Venus there was what may be termed a runaway greenhouse effect, and any life that had appeared there was ruthlessly wiped out.

What role can the Moon have played in the story of the youthful Earth? Sir George Darwin (son of Charles Darwin) suggested in 1878 that the Earth and the Moon used to be a single body which was in rapid rotation; the whole mass became pear-shaped and then dumbbell-shaped until the 'neck' of the dumbbell broke and the Moon moved away independently. Later it was even suggested that the great depression of the Pacific Ocean marks the place from which the Moon departed. Actually, all this seems improbable. The Moon could not have been thrown off in such a way and moreover the Pacific Ocean is relatively shallow, while the Moon has more than one-quarter the diameter of the Earth. It is now believed either that the Moon was formerly an independent planet, or else was produced in the same region as the Earth and at the same time, so that the two bodies have always remained gravitationally linked. It is even possible that the Moon accreted from material which formerly orbited the Earth. In any case, it is quite definite that the Earth and the Moon are of the same age; analyses of the lunar rocks brought home by the Apollo astronauts and the Russian automatic Luna probes have shown that.

If modern theories are correct, it follows that solar systems must be relatively common in the Galaxy; what can happen to the Sun can happen to other stars too, and there is every reason to assume that just as the Sun is a normal star, so the Earth is a normal planet. In 1983 the IRAS infra-red satellite detected material around two stars, Vega and Fomalhaut, which may well represent planetary systems in the process of formation. But whether the Earth is exceptional in supporting intelligent life is quite another matter. For the moment, at least, we have no proof of any extra-terrestrial life (even on Mars), and we may be sure that there are no other advanced beings in the solar system. It is fortunate for us that the Earth was born at just the right distance from the Sun, and was of just the right size and mass. If things had been even slightly different, you and I would not be here.

3

THE PRIMAEVAL EARTH

Birth of a Planet

There may be heaven; there must be hell;
Meantime, there is our earth here – well!

Robert Browning: 'Time's Revenges'

Our conception of the Earth's birth has changed many times. Even during the 'space age', which began approximately when the Russians launched their first successful Sputnik in 1957, our ideas on this topic have been in a state of flux. Intensive studies of meteorites – pieces of rock which date back to the very beginnings of the solar system, and which in some cases have avoided modification ever since – have led us somewhat nearer to what may reasonably be considered the truth. Even so, we must exercise some caution, as there are many unknowns, even in the most modern hypotheses.

There is little room for a reasoned belief that the Earth was suddenly conjured up from some universal 'stuff', at a specific moment in time. Our knowledge of the chemical and physical processes that operate within the universe has taken many strides forward since the days of Newton. Thus we are now fortunate to have samples from the Moon, to have remotely studied rocks on both Venus and Mars, and to know the ages of many meteorites and rocks from both the Earth and Moon. Although we still cannot say exactly how the Earth started to grow, we have a better understanding of the time scale of events and of at least some of the phenomena that undoubtedly played an important part in the Earth's formation and growth.

The modern view is that the Earth, like the other solar system bodies, accumulated rather quickly in a cloud of dust and gas surrounding the primitive Sun. The evidence for such rapid growth is to be found in the decay products of radioactive isotopes of certain chemical elements, such as iodine, plutonium and aluminium, which are found in meteorites. Atoms of these radioactive isotopes disintegrate with time, producing other non-radioactive atoms that remain in the meteorite body. The change to these other isotopes happens very quickly indeed on the cosmological scale, and the fact that a record of these short-lived radioactive isotopes still exists in some meteorites means that the particles containing them must have been very rapidly incorporated within them. This suggests a quick growth for the meteorite bodies.

The meteorites have also retained important chemical evidence relating to what went on during the formative stages of planetary

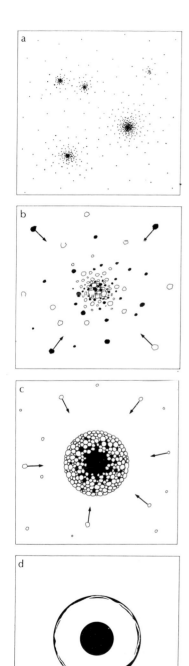

Accretion of the Earth: (a) dust particles form in the nebula; (b) larger particles are attracted to one another; (c) planet-sized body forms with heavy elements sinking to form a dense core; (d) the outer mantle builds up.

growth. Radiometric dating of the oldest of the meteorite groups, called chondrites, gives an age of about 4600 million years. This is also taken to be close to the actual age of the solar system itself. Thus the small bodies (called planetesimals) from which the meteorite fragments derived must have accreted, melted, differentiated chemically and cooled all within a few million years.

Little factual evidence exists concerning how accretion actually occurred. It is generally believed, however, that at some point nuclei grew within the cloud, and that from these the planets gradually grew. In the case of the Earth, it is thought that growth began with dust grains which, with the assistance of weak electrostatic forces, were gradually converted into centimetre-sized particles. These gradually aggregated into larger and larger bodies until first asteroidal, and then planetary-sized bodies evolved. This early phase must have been very complex; probably periods of growth alternated with phases of disruption and then reaggregation, this occurring many, many times before growth was completed.

All of the larger solid bodies within the Sun's family were subjected to intensive bombardment by meteorites during the first 500 million years of their existence. The surfaces of all are heavily cratered, each crater being the scar left after a meteorite impacted on the surface. It was probably the constant impacting of particle against particle that gradually built the planets, their satellites and the asteroids.

The kinetic energy of the impact process is ultimately converted into heat; thus, as a primitive planet grows, so it attracts larger and larger planetesimals, these impacting at higher and higher velocities. Eventually substantial amounts of kinetic energy are transformed into thermal energy, for which reason it is probable that the outer layers of a primitive planet would get hotter than the interior. Much of the heat would, of course, be radiated away into space, but where very massive impacts occurred, some thermal energy would have been implanted deep inside a planet like Earth; this would not escape easily.

In the cases of both Earth and Venus, which have substantial central regions, called cores, of heavy elements like iron and nickel and which are considered to be at very high temperatures, the actual formation of such cores would in itself have been an important source of internal heat. Thus the downward movement of droplets of heavy, core-forming materials under the influence of gravity would actually generate heat.

Gradually, then, the Earth would have become hot and, from what we have learned about the Moon in recent years, large parts of it would have actually melted, producing 'magma oceans' – huge volumes of molten rock, like volcanic lavas. These would have formed the outer layers of such a planet as Earth. Gradually these would have cooled, whereupon there would have been a slow thickening of the solid layer at the exterior of the planet, which we know as the crust.

The Early Atmosphere

As the early planets began to evolve, so too did the proto-Sun. The inward pressure of gases within the solar nebula eventually reached such high levels that thermonuclear reactions were triggered off. At this time the Sun was truly 'born'. Many young stars have been studied, and it has been shown that at this stage in a star's history a large amount of gas is blasted off into space. This is the solar wind. Its importance was discovered when spacecraft studied it in the 1960s.

The outward-moving solar wind (fast-moving solar particles) stripped away the primitive atmospheres of the terrestrial planets at an early stage.

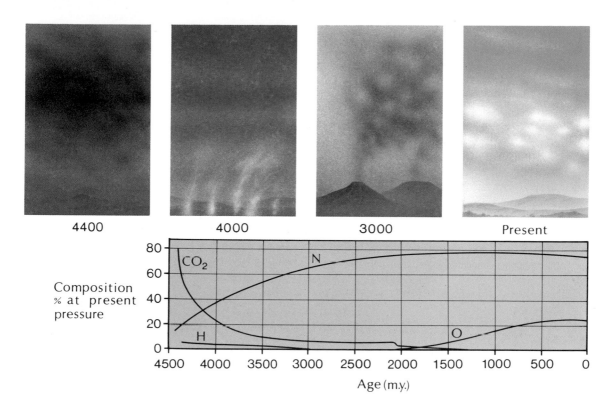

4400	4000	3000	Present

Composition % at present pressure

Age (m.y.)

Schematic diagram to show how the Earth's atmosphere evolved. The decreasing amount of carbon dioxide was balanced by an increase in the nitrogen content. With the rise of plant life, about 2500 m.y. ago, free oxygen began to enter the atmosphere.

The solar wind consists essentially of fast-moving protons and electrons which, by the time they reach the Earth's orbit, are travelling at velocities of about 400 km s^{-1}. Although the solar wind is so tenuous, with a particle density of a few tens of particles per cubic centimetre, its effects are far-reaching. In the case of a newborn star such as our proto-Sun, the wind becomes a veritable gale; this is the T Tauri stage (the name comes from the first star discovered to be at this point in its evolution).

The Earth's primordial atmosphere was stripped off during the Sun's T Tauri stage, and the remaining hydrogen and helium were blasted out of the solar system into interstellar space. This cosmic cleaning-up process saw the stripping away of the atmosphere of all the inner planets, so that only their solid parts remained. Therefore only the planets, asteroids and meteoroids survived this first 'creation' phase, and most of the mass of the solar system (99.8 per cent) has remained in the Sun.

The Earth's present atmosphere is entirely secondary, and was derived from the hot interior of the globe by a process known as degassing, which occurred as the mantle layer solidified. By the time that this took place, much of the metallic iron had sunk toward the core, and oxidizing conditions promoted by the relatively oxidized mantle resulted in the production of oxidized gaseous forms of nitrogen, sulphur and carbon (i.e. NO_2, SO_2, CO_2). Confirmation of this secondary origin is to be found in the relatively low abundance of the noble gases neon, krypton, and xenon on the Earth as compared with their abundance in the Sun.

Further modification of our atmosphere has taken place because of biological processes such as photosynthesis. Over the last 3000 million years, living organisms have relentlessly changed the atmosphere in such a way that our present air is quite different from the atmosphere of Venus or Mars: planets that in many other ways are not too unlike the Earth. The Earth differs also in being largely covered

19

with water; over 70 per cent of our globe is covered by oceans.

However, Venus, the Earth and Mars can be treated as a group when compared with the outer planets. Whereas the terrestrial planets have atmospheres that hug their surfaces and are relatively scanty, the planets of the outer group have gaseous envelopes tens of thousands of kilometres deep. This striking difference is the direct result of the differing compositions of primordial grains inside the solar nebula.

The Great Bombardment

Planetary surfaces are moulded by two main groups of processes: endogenic (such as volcanicity, mantle movements and tectonics) and exogenic (specifically, bombardment by meteoroids). In the past, traditional geology has confined itself to studying the first of these groups, but gradually it has become clear that the role of the second is of tremendous importance. This realization resulted from the exploration of the Moon, during which scientists came to appreciate the effects of impacts upon planetary surfaces.

Today, the solar system consists of a relatively orderly collection of larger objects – the planets, their satellites and the Sun; but there are also somewhat less well-ordered bodies – the asteroids, comets and meteoroids in particular. In very early times the content of disorderly material was far higher than it is now, and collisions between the proto-planets and this loose debris were commonplace. Since a collision between a planet and a fast-moving fragment is a very energetic event, considerable damage will be done to both.

Spacecraft images of Mercury, the Moon and Mars and, indeed, many of the satellites of Jupiter and Saturn show that their surfaces are pockmarked with craters, both large and small. Most of these were produced by collisions with wandering meteoroids during a period of intense bombardment which took place over 4000 million years ago. This phase of early impacts has often been termed the Heavy or Great Bombardment.

The amount of energy released during a fairly typical impact can be appreciated from the following figures. Consider a meteoroid with a relative velocity of 16 km s^{-1}. The kinetic energy of such an object impacting the Earth is 1.3×10^{12} erg g^{-1}. The bare figures may not mean much, but when we compare them with the chemical energy of, say, TNT, which works out at roughly 4.2×10^{10} erg g^{-1}, its potential for destruction becomes more readily apparent. Unless such a body were so decelerated and abraded during its fall through the Earth's atmosphere that it lost virtually all its mass (which would occur if the body were very small), its impact on the Earth's surface would blast out a crater with a volume thousands of times larger than that of the original meteoroid. The impacting body would disintegrate, and would be dispersed along with the terrestrial debris from the crater.

No evidence of major impact events has been left in the Earth's oldest rocks. However, this does not imply that no such impacts occurred. About 70 impact craters have been partially preserved. These are often vague remnants produced by erosion of the shattered rocks in the roots of the craters, but others are relatively young and are better preserved. Barringer Crater in Arizona (~0.35 million years old) is a case in point. Because the old continental shield areas, such as those in Canada, Africa and Australia, are relatively stable, have suffered relatively little vertical movement and have been least affected by erosion, most of the older impact craters preserved are to be

Map showing the distribution of supposed impact craters on the surface of the Earth. Not all geologists would agree with this interpretation.

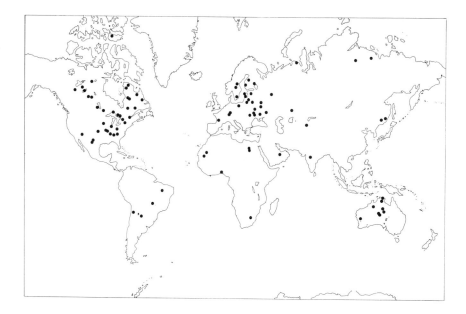

found in these ancient terrains. The rarity of such craters when compared with the Moon or Mercury is due in part to the comparatively rapid rate of weathering and erosion on the Earth, which will remove a crater quite quickly on the geological time scale, and in part to the restless nature of much of the Earth's crust.

Most geologists would subscribe to the view that the Great Bombardment really did affect the Earth, but experts in the field of lithospheric development will probably argue that the Earth was too mobile for it to have had a primitive surface akin to that of the Moon. What is perhaps most interesting is that the earliest life had appeared by about 3500 million years ago, just after the period of heaviest bombardment ended. Did the close of this phase make way for a very different Earth? Well, there is certainly strong evidence to suggest that the volatiles necessary for the atmosphere and hydrosphere were present 3500 million years ago, perhaps earlier, and that these formed the basis for organic development. Some workers have suggested that these volatile materials were brought to the Earth by the meteoroids themselves. If this were so, then it would clearly be true to say that the close of the bombardment marked the beginning of a new epoch in Earth history.

The Primitive Earth

As we have seen, the difference between the inner planets and the outer group is due to the temperature difference inside the primordial nebula from which the planets were formed. Temperature had a profound effect upon exactly what happened inside the dust-and-gas cloud, because the various kinds of matter in the solar system have rather different chemical characteristics.

First there is rocky material, composed largely of silicates and metal oxides, chiefly of magnesium, aluminium and iron. Next there are ices and liquids, which are combinations of carbon, nitrogen, hydrogen and oxygen (such as water, H_2O). Finally there are those substances that, except under very unusual conditions, are gases – such as hydrogen, helium, neon and argon. Of these three groups, the first is characterized by very high melting points, around 1250°C; materials of the second group have much lower melting temperatures of around

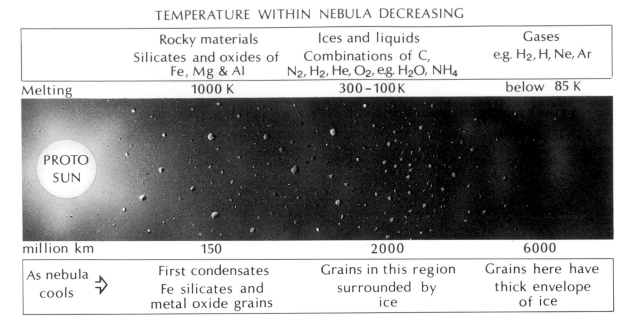

	Rocky materials Silicates and oxides of Fe, Mg & Al	Ices and liquids Combinations of C, N_2, H_2, He, O_2, e.g. H_2O, NH_4	Gases e.g. H_2, H, Ne, Ar
Melting	1000 K	300 – 100 K	below 85 K
million km	150	2000	6000
As nebula cools →	First condensates Fe silicates and metal oxide grains	Grains in this region surrounded by ice	Grains here have thick envelope of ice

370 to 570°C; those of the third group remain gaseous except under conditions of extremely low temperatures.

As material plummeted toward the centre of the nebula during its early history, the temperatures rose to several thousands of degrees C. In these more central regions all matter was vaporized. In the outer parts of the cloud the temperatures were much lower, probably never exceeding 370°C, so that in these regions the dust and rock grains were coated with dry ice, water ice, and frozen methane or ammonia.

Gradually, as the proto-Sun evolved, the temperatures inside the nebula declined, so that gases cooled and solid material began to condense out. Because of their high melting points, the rocky materials naturally condensed first, but near the centre of the cloud it was only the rocky substances that could solidify. In these inner regions the solid grains were largely of iron, silicates and metal oxides. Further out, thin ice layers surrounded the dust grains, and at still greater distances, say the distances of Jupiter or Saturn, the grains were thickly enveloped in icy coatings. A common feature of all the grains, whether close to or far from the proto-Sun, was that they were enveloped in a huge cloud of hydrogen and helium which accounted for over 95 per cent of the volume of the nebula itself.

When the Earth's mass had grown to about one-tenth of its present value, gravitational attraction became strong enough to make the incoming planetesimals strike the proto-Earth at very high velocities – high enough, in fact, to vaporize them on impact. When this happened, as it did on millions and millions of separate occasions, volatiles were released, and these went to form a primitive atmosphere. Non-volatile elements produced by the same process were condensed, and gradually built up an outer layer of molten rock. As a result, the early Earth was probably made up of a cool volatile-rich interior surrounded by a thick ocean of molten rock, perhaps 1000 km deep.

With the passing of time, the decay of radioactive isotopes locked inside the inner planets gradually heated and melted their interiors. Under the influence of gravity, the denser materials sank toward their centres, causing chemical differentiation, so that each planet developed an iron-rich core surrounded by layers of silicate rocks.

The effect of temperature upon events within the primordial nebula. This had profound repercussions for the distribution of rocky, icy and gaseous materials in the solar system.

4

METHODS OF THE EARTH SCIENTIST

Early Ideas and Methods

While it may be true to say that modern geology did not develop until the late eighteenth and early nineteenth centuries, the early pioneers were fortunate in having at their disposal a vast accumulation of observations, speculations and philosophical writing that had been amassed during earlier times. The first faltering steps into the then unknown world of astronomy were taken by the Egyptians during their

Geological map of Oxfordshire, England, by William Smith. This map, laid out by John Cary in 1821, is typical of Smith's superb craft. The original scale was 2.5 miles to the inch.

first dynasty, about 3000 BC, but it was left to the Greeks to take the initiative and explain some of the perplexing features of the Earth.

Before the time of Pythagoras, about 530 BC, most natural events were attributed to the activities of gods or mythical heroes. There came a time, however, when superstition and primitive mythological explanations gave way to more carefully reasoned arguments. Admittedly, the seeds of truth which some of the early theories contained were entirely coincidental, but even so we owe them much for the attempts they made to understand the natural processes that were an everyday part of their lives.

The Mediterranean had long been seismically active and there are many accounts of earthquakes in the Greek records. Furthermore, the southern Aegean and the south of Italy and Sicily, where the Greeks had colonies, are volcanically active: witness the volcanic caldera of Thíra (Santorin) and the chain of volcanoes that extends from Naples, past Vesuvius to the Lipari Islands, Etna and beyond. From the dawn of classical times, people must have lived in awe of these fearsome mountains and the trembling earth beneath their feet. Within the Mediterranean region there is also a remarkable diversity of climate and a consequent complexity of geological processes over which climate is the control. There are high mountains, such as the Alps and Apennines, glaciers and snowfields, countless fast-flowing rivers, and huge deposits of rock debris that have been flushed from regions far inland by these rivers, and spread out along the coastline of the Mediterranean Sea.

To follow the development of Greek ideas about geology would take many pages and there is not space here to pursue this fascinating topic. Nevertheless, we must make note of the fact that it was the Greeks who first tumbled to the real nature of fossils, connected volcanic activity with earthquakes, and realized that areas now high up in the mountains were once beneath the sea. Not all thinking persons held such views, however, and very many still believed fossils to be 'sports of nature' – a view that was to survive through medieval times. The change towards more modern ideas was a very slow one.

The English antiquary Joshua Childrey appears to have been the first writer to refer to fossils (*Britannia Baconica*, 1660), and Robert Hooke, in his famous work *Micrographia*, published in 1665, produced fine drawings of fossil shells and careful descriptions of them. Buffon, with his *Histoire Naturelle*, published in 44 volumes, 1749-1804, and in particular the volume *Epoques de la Nature* (1779), also deserves mention. In the latter work we have the first serious attempt to compute the age of the Earth. Jean Etienne Guettard (1715-86) was undoubtedly the first of the great mineralogists and he also had a profound knowledge of fossils. His mineralogical map of France and Britain contains a wealth of accurate information, based on painstaking field research.

It was not until 1788 that the next truly major step was taken. This step was marked by the publication of James Hutton's paper, 'The Theory of the Earth', in the first volume of the *Transactions of the Royal Society of Edinburgh*. In this, Hutton expounded his concept of geological cycles, during which similar processes had operated over very long periods, in much the same way as now. This became known as the 'Principle of Uniformitarianism', and marked the beginning of modern geology. In 1795 Hutton published this paper, and a host of other writings, in a two-volume book that has become one of the classics of geological literature: this was called *Theory of the Earth with Proofs and Illustrations*.

Mapping Rocks

As far as we can tell, the very first piece of geological 'mapping' was George Owen's description, in words only, of the outcrop of the strata known as the Carboniferous Limestone, around the South Wales coalfield. This was published in 1603. George Sinclair, who in Scotland in 1672 published a volume entitled *Hydrostaticks*, seems to have given the first detailed account of a geological structure. He explains the meaning of 'dipp' (dip), 'streek' (strike) and 'crop' (outcrop) and also gives a clear description of the folding and faulting of rock strata.

The term outcrop refers to any exposure of bare rock. If this is a sedimentary rock – such as a marine sandstone, composed of derived rock fragments laid down under water, then buried, compressed and

finally brought to the surface again – it will probably be stratified. By this we mean that it will have been laid down as a horizontal or near horizontal layer of grains which subsequently became coherent, giving rise to a bed or stratum. The study of stratified rocks forms the basis of geological mapping. Analysis of any fossil remains such a bed contains allows geologists to put the geological history of an area into the broader context of Earth History.

Strata may not, of course, remain horizontal. The Earth is a dynamic planet, and periods of internal upheaval may deform rocks, rucking them into folds, tilting them and even fracturing them. Folded strata are very common and if deformation is very intense, it may cause them to be turned upside-down. This produces a very complex situation which only detailed mapping of the rocks will elucidate. A geological map tries to show three-dimensional structures in rocks on a two-dimensional piece of paper. Thus, where beds are inclined due to folding or tilting, the map will show the angle at which they are inclined to the horizontal; this is known as the dip of the beds. The trace of a line drawn perpendicular to the dip is termed the strike of the beds. Dip, strike and outcrop are three fundamental geological terms.

Rock outcrops, also called exposures, may well be widely scattered over a region. Between them may be soil, vegetation or urban developments; in some instances, seas or oceans will separate outcrops of similar rocks. One of the tasks of the geologist is to correlate beds lying in widely separated areas. To do this he or she needs not only to study the nature of the rocks themselves, but also any fossils they contain. Fossils are the remains of past organisms and it is now well understood that organic life has evolved in particular ways. Thus, certain kinds of organisms lived only during particular intervals of time, while specific genera or species evolved in particular ways. Palaeontologists, that is geologists specializing in the study of organic remains found in strata, now have a pretty good idea of the different kinds of organisms that lived in the many kilometres of strata that lie on the Earth's crust; they know the sequence of rocks and the particular fossil faunas which rocks of different ages contain. This is why mapping a region involves both the charting of the rock strata and the inspection of any fossils within them.

Today, there are geological maps of most regions of the Earth, even remote tracts such as the Siberian plain and Antarctica. There are even very incomplete charts of the rocks beneath the oceanic waters, as well as on the surface of the other terrestrial planets. Modern technological developments could not possibly have been suspected by the early pioneers. How could they have predicted that orbiting satellites would now be making such a major contribution to our knowledge of the Earth, not only of its rocks, but also of its waters, weather and vegetation? It is the development of rockets and the spacecraft they may carry as their payloads that has enabled scientists to learn about the rocks of other worlds, like the Moon and Venus. It is with great interest that we await the construction of orbiting space stations that may form the springboard from which the next decade's most spectacular developments will be seen to start.

Returning to more basic matters for a moment, perhaps the finest of the early geological maps were produced during the latter part of the eighteenth century, in particular by the famous engineer-cum-surveyor-cum-geologist, William Smith. He was a mineral surveyor who, in his travels, visited numerous mines and quarries. During his

Looking down the 'strike' of a series of almost vertical Silurian strata, Aberystwyth, Wales.

work he began to recognize the regularity of the strata he encountered, and the great value of the fossils they contained. When subsequently he became involved in the surveying for the great canal network that was built in southern Britain in the late eighteenth century, he came quickly to recognize the different strata in widely scattered regions, not only because of their colour or mineral content, but also by the fossils they contained.

Between 1794 and 1809, Smith collated a series of geological sections across various parts of England. His immaculately drawn section of the strata between Bristol and Norwich appears to have been the first of its kind. Even more impressive were his fine geological maps, particularly the one published in 1799 and titled: 'Geological Map of the country around Bath'. By 1815, he had published a series of fifteen maps that covered most of England and Wales. These are now classics. His 1801 map was the first true representation of the geology of England and Wales on a single sheet. We owe much to this great man, not only because he achieved so much, but also because he set such a high standard for later workers to emulate.

The Fossil Record – 1

Most ancient writers variously explained the preservation of fossils in rocks as the result of great natural catastrophes which effected a change in the relative positions of land and sea. With the dawn of the fifteenth century, and the revival in learning that accompanied this, came three centuries of debate: were fossils 'sports of nature', or did they originate in some kind of magical fluid emanating from within the Earth (called at the time *vis plastica*), or were they indeed the remains of past life forms?

The fifteenth-century genius Leonardo da Vinci was among the first to realize that fossilized remains contained in sedimentary strata meant that the rocks themselves had once been beneath the sea. In the latter part of the sixteenth century the Frenchman Bernard Palissey (*c.* 1510-89) published drawings and descriptions of petrified wood, together with fossilized fishes and molluscs, rightly concluding that these occurred in rocks that represented sediments which once had been submerged by water. He was verbally attacked by his contemporaries and soundly denounced by the Church. The English zoologist Martin Lister (1638-1711) had an excellent knowledge of

Early drawings of marine fossils. This particular set appeared in Johann Jacob Baier's *Oryctographiae Noricae*, pubished in 1712.

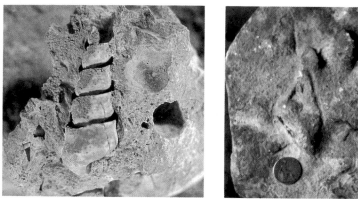

What the palaeontologist refers to as body fossils are the relatively tough skeletal remains which survive intact after the soft tissues have decayed. Such are the shells of molluscs and crustaceans and the bones of vertebrates. Even these, however, may come under attack; thus marine shells may be bored into by other organisms or broken up while being transported across the ocean floor by currents, or even dissolved in chemical reactions occurring after burial.

Those parts that survive such rigours must eventually become buried by sediment if they are to survive further degradation, and therefore animals that live within the sediment itself are at a distinct advantage. They have a substantially greater preservation rate than, say, free-swimming forms or bottom-crawlers. The survival of plant fossils is an equally chancy business. Very delicate vegetable remains may be preserved if they are buried quickly, for example by sand or mud brought down into lakes or low-lying areas by earthquakes, volcanoes or other catastrophic phenomena.

After burial organic remains will be subjected to compaction as the overburden gradually increases in depth. Some kinds of skeletons will be crushed during this process, others will survive reasonably well. In addition, replacement or chemical alteration of the original organic material may occur. This is very common among calcareous skeletons.

Calcareous organisms are composed of calcium carbonate. This occurs principally in two forms, calcite and aragonite, which differ in their internal constitution. The latter is particularly unstable and usually is dissolved away, or is converted to the more stable form calcite. Aragonite thus becomes less often encountered as we move down through the rock succession to older horizons. Sometimes the calcite or aragonite may be totally removed in solution, which leaves a cavity. Subsequently this may be filled by more stable minerals. Such a process gives rise to what is termed a mould or cast. Such modes of preservation frequently reveal delicate anatomical features that otherwise would not be visible. At other times the calcium carbonate may simply be replaced, crystal for crystal, by silica or iron pyrites.

A relatively small proportion of animals actually have a skeleton composed of silica: examples are the sponges. Others may be composed in part of calcium phosphate, like the conodonts, which are preserved in apatite, a mineral that is quite resistant to degradation. Whatever the composition of the skeletal material, however, this has a substantially greater chance of survival than the soft tissue. Nevertheless, occasionally these delicate tissues are preserved, and such a discovery has often made the press headlines. One example was the discovery in Siberia of a frozen baby mammoth that had lived at least 10,000 years ago.

Another kind of fossil is the trace fossil. This represents a preserved track or other markings made in sea-floor sediment by organisms which have been active in the surface or subsurface sand and mud. Usually it is impossible to establish which particular organism produced the impressions, but in certain instances more definite identification can be made; thus traces of activity by trilobites have been well documented from ancient rocks of Early Palaeozoic age in Wales and many other parts of the world. The more common trace fossils represent activities like crawling, burrowing, feeding or resting. The majority appear to have been made by animals actually living in the sea-floor sediment. Their greatest value is as indicators of water oxygenation and diversity of the original biota (the flora and fauna of a region); they have some value as indicators of the environment in which the organisms lived.

Fossil plants are frequently found in strata associated with coal deposits, and although they are usually severely compacted and the internal features destroyed, they are often beautifully preserved as impressions. Occasionally the internal anatomy is revealed because the plant was invaded by mineral solutions before it could be compressed and crushed. The famous silicified forests of Arizona are examples of this kind of preservation.

Although our knowledge of the fossil record is substantially more complete than in the days of Smith or Lamarck, there are still large gaps. Some of these are due to the almost complete destruction of some delicate types of organism; others remain because palaeontologists have yet to study and collect from all available localities.

Methods of the Geophysicist – 1

In their quest to perceive what lies at greater and greater distances from the Earth, astronomers have invented successively more powerful telescopes which have exploited different parts of the electromagnetic spectrum. The geophysicist has pursued a similar course in an attempt to penetrate down into a planet that is not transparent to light waves. Fortunately the Earth's rocks allow passage to sound or 'seismic' waves and the recording of such vibrations has formed the basis not

a PENDULUM – MOUNTED SEISMOMETER

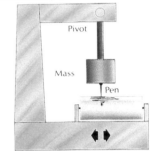

b SPRING – MOUNTED SEISMOMETER

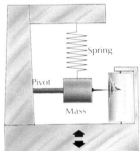

ABOVE Seismographs:
(a) pendulum-mounted instrument that records horizontal movements;
(b) spring-mounted instrument that records vertical motion.

BELOW Seismic reflection profile across part of the North Atlantic Ocean. The subsurface structure and sedimentary layering are clearly revealed in this traverse.

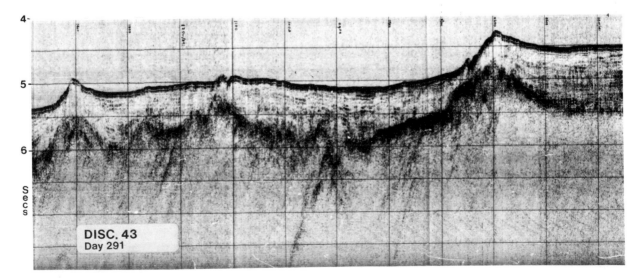

only of our understanding of the Earth's internal structure, but also of several commonly used methods of prospecting for oil and gas.

Earthquakes generate seismic waves which travel through rocks and whose nature and arrival time can be accurately recorded at the surface on delicate instruments called seismographs. These were first invented in the late nineteenth century, whereupon scientists confirmed not only the existence of compressional and shear waves but also that the two kinds of seismic vibrations were not transmitted with equal facility through all parts of the Earth. Interpretation of these early findings led to our first reliable ideas about the Earth's internal structure (see Chapter 6).

A seismograph is a device for measuring any change in distance between itself and the Earth beneath it. Theoretically the device is not itself affected by movements in the Earth, but in reality this cannot be achieved. Instead an approximation to this condition is realized by mounting a small mass on either a pendulum or the end of a delicate spring, so that when the ground moves the mass remains still because of its inertia, while the Earth is displaced relative to it. The displacement is recorded by the seismometer.

Modern exploration geologists generate their own mini-earthquakes to gain information about the subsurface strata, thereby increasing the efficiency with which they can predict the presence of oil and gas. A series of small seismic waves is generated, sent down into the crust, reflected by the subsurface rocks and then arrives back at the surface of the Earth, to be recorded by a detector. Generally a series of detectors is used, these being arranged in a straight line which stretches away from

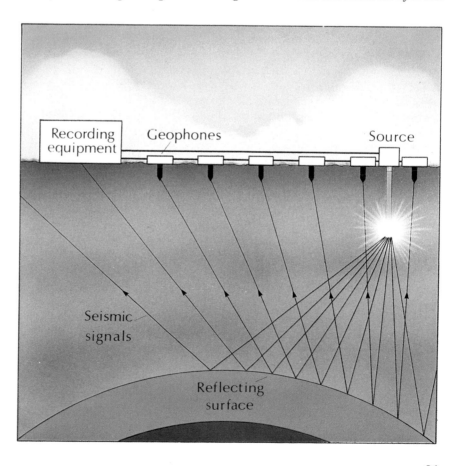

Seismic technique for probing subsurface geology. Small explosions generate sound waves which are bounced off rock layers beneath the surface and bounced back to the geophones.

the seismic source. On land the energy source is sometimes some kind of explosive. This method is sometimes also used under water, but it is more common to employ a special kind of air gun.

To record the reflected waves, the large seismometers necessary for earthquake work are not required. Instead small recorders called geophones (for land work) or hydrophones (for marine work) are used. During marine surveys these are usually towed behind a research vessel in a line or 'seismic streamer', at a depth of between 5 and 15 m of water. Computers process the signals produced, apply certain corrections, and issue a seismic reflection profile. Such a profile reveals not only the subsurface structure but also gives information about the physical make-up of the hidden rocks.

Another physical method utilizes a logging tool called a sonde, which is lowered on a steel cable into a bore hole, and which makes a variety of measurements aimed principally at establishing the porosity and permeability of the strata encountered. This Electrical Wireline Logging technique is widely used in the oil and gas industry, where decisions to exploit or not to exploit a new field are based on the well 'logs' so obtained. The sonde itself may contain instruments to make a variety of electrical measurements, and probably readings of natural gamma ray radioactivity. Together these can provide vital information on rock porosity, the ease with which fluids can travel through the rock pores (permeability) and the nature of the fluids trapped within the pores themselves. The log is made as the sonde is gradually withdrawn from the bore hole at a carefully controlled rate, the results being plotted as a curve of measurements versus depth. Also within the sonde are the electronics to power the signal transmission and send data back to the recording instruments on the surface.

Methods of the Geophysicist – 2

About a century and a half ago, British surveyors involved in the mapping of the Indian subcontinent encountered an inconsistency in their survey which led to the discovery of one of geology's fundamental concepts, that of isostasy. The details of this discovery are sufficiently interesting to describe in a little detail. Briefly, the distance between Kaliana, about 150 km south of the Himalayas, and Kalianpur, some 600 km further south, was determined in two very precise ways: by measurement over the land surface and by reference to astronomical observation. The results disagreed by some 150 m, which may appear trivial, but to the surveyors was an unacceptably large discrepancy. In trying to explain away the difference, it was proposed that the plumbline used in the survey was tilted towards the mountains because of the gravitational attraction they had on the plumb bob. Calculations of the theoretical effect indicated the discrepancy should have been even greater than it appeared to be – 450 m! Thus the problem was compounded rather than solved.

In 1865 Sir George Airey, then the Astronomer Royal, came forward with the suggestion that the enormously heavy Himalayas were not actually a part of a rigid crust but were buoyed up on a layer of denser rock, rather as the hull of a ship in water. In effect this meant that the excess mass of the mountain chain above sea level was compensated for by a mass deficiency in the underlying 'root' zone. This explanation became the accepted one and formed the basis of the geological concept of isostasy, which has played an extremely important part in our understanding of the behaviour of the crustal rocks.

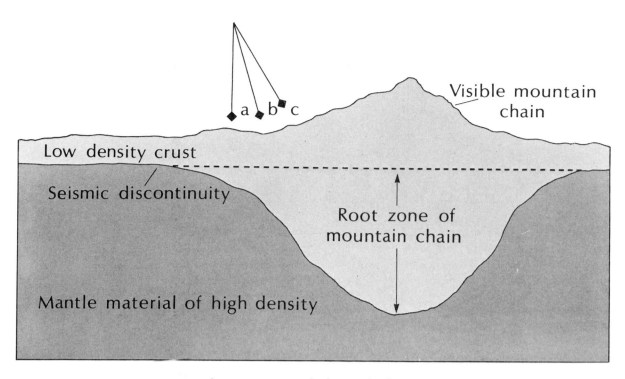

Isostacy. Mountain masses deflect a pendulum away from the vertical, but not as much as might be expected. In the diagram the vertical position is shown by (a); if the mountain were simply a load resting on a uniform crust, it ought to be deflected to (c). However, because it has a deep 'root' of relatively light rocks, the observed deflection is only to (b).

The same principle forms the basis of another common method of Earth surveying, that of the gravity survey. The objectives of this are to measure the mass variations that occur inside the Earth and they are achieved with the use of an instrument called a gravimeter. This is a small device, usually no larger than a wine bottle, and consists of a weight on a spring that stretches or contracts in response to either an increase or decrease in gravity from place to place. Modern instruments can detect variations as small as one hundred millionth of the Earth's gravity. The standard unit of measurement is the milligal which is equal to a gravitational acceleration of 0.001 cm per second per second.

As with many of the modern geophysical techniques, costs involved in development have been borne mainly by the oil industry which, about 40 years ago, realized that the buried geological structures in which oil is trapped often produced variations in the normal gravity field, and that these could be detected by sufficiently sensitive recorders. Such gravity anomalies are caused by a change in the subsurface mass due to the presence of such structures as salt domes.

The philosophy behind such surveys is that any anomalous mass will have an effect upon the gravitational field and that careful surveying of the gravity may be used to reveal details of the subsurface geology. Before a profile can be useful, however, a number of corrections need to be made and these arise largely because the Earth is not a perfect sphere. The polar flattening, produced by the centrifugal forces that operate in the rapidly rotating planet, means that the force of gravity is less at the Equator than at the poles. Thus a correction must be made for latitude. Furthermore, a gravity station on a hill is further from the Earth's centre than one at sea level and will therefore give a lesser measurement. This effect must be taken into account. Lastly, if the profile eventually derived is to be properly representative of the subsurface geology, care must be taken in assessing the gravity contribution of all near-surface masses of rock, correcting accordingly. Having made these very necessary corrections, gravity surveys become a very potent tool in the elucidation of the Earth's subsurface geology.

Part Two

THE PLANETARY
ENGINE

5

THE EARTH'S HEAT ENGINE

The Earth Warms Up

By making very reasonable guesses about the early temperatures and pressures inside the Earth and also about the thermal contribution of radioactivity, it is possible to compute what may have occurred during the first billion (a thousand million) years of Earth's history. All such calculations indicate that accretional and radioactive heat was unable to escape fast enough to prevent the Earth heating up. Thus by 3500 million years ago the internal temperature was high enough to cause the melting of metallic iron; this had a fundamental effect on the Earth's subsequent evolution.

Many factors suggest that the Earth began accreting after the metallic and silicate grains had condensed from the solar nebula. It seems that metallic particles, containing iron, were preferentially accreted, due probably to the greater ductility and higher density of iron-rich molecules compared to the more brittle silicates that subsequently accumulated to form Earth's mantle layer. Evidence provided by meteorites suggests that in addition to small grains, there were larger planetesimals, at least 100 km in diameter, that existed very early on and that such bodies also contributed to the Earth's construction. This being so, it was not possible to delay for long the inevitable formation of a dense planetary core, a process that has been shown to have occurred during the early evolution of all of the terrestrial planets.

As the planet grew, so gravity increased. Calculations indicate that the sinking of iron could have started after only one eighth of the Earth's mass had accumulated. Thus it seems that core formation started well before the whole Earth had accreted; once the separation of iron and silicates began – a process called differentiation – Earth would have heated up much more rapidly as gravitational energy was released. If accretion was very rapid, then the radiation of thermal energy into space would have been exceedingly slow and heating-up would have been even faster. Certainly there is every reason to believe that during this phase the Earth's internal temperature rose by around 2000°C. This would have triggered wholesale melting and set in motion an even more efficient process for redistributing Earth's chemical ingredients: convection.

Stirrings Inside a Planet

The formation of the core marked the beginning of a new phase in Earth's history. It became a layered body in which the denser components sank toward the centre while the lighter materials accumulated nearer to the surface. Estimates for the time of core formation range between 4.5 and 3.7 billion years ago. At this stage the internal heat of the Earth could escape only by conduction – a very

Cerro Negra volcano, Nicaragua, erupting in 1968.

slow process – so the internal temperature rose greatly until much, if not all, of the Earth became molten.

At this point convection, a more efficient mechanism for the transfer of thermal energy, took over. Any body of liquid convects if it is hotter at the bottom than the top: the hotter material expands and floats upward over the denser, cooler material above until it, in turn, cools, becomes more dense and then sinks again. By this process heat was more efficiently transferred toward the Earth's surface and dissipated away into space.

In time the Earth cooled sufficiently for its outer regions to solidify, although the core remained molten, as the outer part of it does until this day. Apparently even 4500 million years is too short a period for the core to have 'frozen'. An analysis of how seismic waves are affected as they pass through the body of the Earth indicates that today the Earth consists of an inner core which is solid and an outer core which is liquid. Together these two layers extend from about 2900 km below the surface to the Earth's centre. Surrounding the core is a less dense solid layer of silicates – the mantle – which lies between the metallic core and the thin, silicate crust, composed of relatively non-dense silicates with low melting points.

Internal convection persists to this day, as is shown by the slow movements of the outer skin of the Earth in response to movements at deeper levels. Yet, how can this be if the mantle is currently solid? Surprisingly enough, under certain conditions, even solid rock can be set into motion. This happens if it becomes very hot, whereupon it expands and becomes less dense. It may then move upward very slowly, giving rise to something akin to a convective motion, called 'high-temperature creep'.

Elemental Abundances

Early in its evolution, the chemical elements residing in the Earth formed compounds that had different melting points, densities and chemical affinities. The subsequent distribution of the chemical elements was a function of the properties of these compounds. For example the feldspars – silicates of aluminium, potassium, sodium and calcium – have low melting temperatures and are of low density. These would be expected to rise toward the surface of the Earth during differentiation, as they did, and became the most common mineral group in the terrestrial crust. The mantle, on the other hand, became a reservoir for the less easily melted, denser silicates of magnesium and iron, like the pyroxenes and olivines.

Of the less abundant elements some heavy ones, like platinum and gold, sank toward the Earth's core. This was not simply because they were dense but because they showed little affinity for either silicon or oxygen and therefore did not enter the lattices of the silicate minerals so typical of the crust and mantle. On the other hand, some heavy elements, such as uranium and thorium, which had affinities for both oxygen and silicon, accumulated in the crustal layer. Surprisingly perhaps, gravity played only a secondary role in determining how the chemical elements were redistributed.

One interesting and significant result of the redistribution was that radioactive elements like thorium, uranium and rubidium were gradually concentrated in the Earth's outermost layers, in complete contrast to their originally even distribution within the Earth. In the early aeons, the decay of these radioactive elements contributed

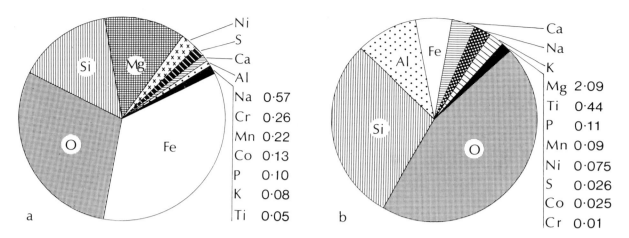

Na	0·57	Mg	2·09
Cr	0·26	Ti	0·44
Mn	0·22	P	0·11
Co	0·13	Mn	0·09
P	0·10	Ni	0·075
K	0·08	S	0·026
Ti	0·05	Co	0·025
		Cr	0·01

Pie digrams showing the elemental proportions in the Earth as a whole (a) and the Earth's crust alone (b).

significantly to raising the temperature of the whole Earth, but as this radioactive fuel started to accumulate in the crust, the heat produced was lost more easily to space. This provides an example of how chemical differentiation had the effect of retarding the action of the 'heat engine' – allowing energy to leak more quickly away.

The Volatiles Escape

The fragments of primitive matter – called 'primeval planetesimals' by many scientists – from which our world formed could not have held on to their volatile elements for long. Because of their relatively small mass they would have had only a weak gravitational pull, and fast-moving gases would have escaped easily into space. The Earth's earliest atmosphere layer must therefore have been produced by gases escaping from the interior. If we accept the view that the Earth accreted from more or less cold fragments, then there is little alternative but to concede that the volatiles that eventually collected on the Earth's surface and formed its atmosphere came from the terrestrial interior. It was the Earth's own internal heat engine that was responsible for their outward motion. This process is called outgassing.

Crustal minerals such as the micas and amphiboles contain, in addition to such elements as silicon, aluminium and iron, both hydrogen and oxygen. These are bound into their lattices as the hydroxyl molecule (OH). As the Earth heated up and chemical differentiation took place, molten rock – magma – was generated by the partial melting of existing solid materials within the planet, and this, together with the water and other volatiles it contained, rose towards the surface. Often it reached the surface, giving rise to volcanic eruptions, and the volatile elements escaped, forming clouds of hot vapour. Calculations based on the assumption that the past frequency of eruptions was similar to today's show that lavas reaching the surface in the past would easily have contributed enough water to fill the ocean basins during geological time. In fact, it is very likely that such eruptions would have been far more frequent than they are today.

The primeval atmosphere must have been quite unlike today's. Modern volcanoes exude gases in large volumes and studies of these suggest that in the early days water vapour, carbon dioxide, carbon monoxide, nitrogen, hydrogen chloride and hydrogen were the most abundant. Of these the very light gas, hydrogen, must have escaped into space at a very early stage; and some of the water vapour in the atmosphere's upper layers would undoubtedly have been broken down into hydrogen and oxygen by the action of sunlight, the former escaping, the latter mostly combining with gases like methane and carbon monoxide to form water and carbon dioxide. The production of large quantities of free oxygen, one of the main components of our modern atmosphere, was as yet a long way off. At this stage the Earth was not an attractive place for the appearance of life.

Volcanicity – 1

The heat engine, fuelled early in the Earth's life by impact energy and the decay of short-lived radioactive isotopes, and subsequently run by the slow disintegration of long-lived radioactive materials such as thorium, uranium and potassium, constantly causes modifications to the Earth's surface. The cycle of heat production, its transfer to the surface, followed by its eventual loss to space, produced fundamental changes in the surfaces of all of the terrestrial planets, and is still doing so on Earth and Venus.

Molten materials are able to flow and it is the differential motion between the more mobile matter deep within the Earth and the comparatively immobile near-surface layers that causes deformation at the surface. In the cases of both Mercury and the Moon it seems that internal heat was lost at a high rate early on in their evolution, thus their surfaces were soon protected from internal modification by thick mantles and crusts which are now essentially cold. Venus, on the other

Volcanic gases escaping from a fumarole in Iceland.

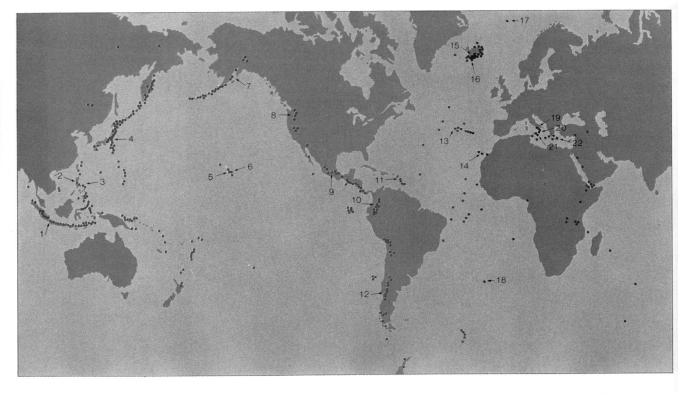

1 Krakatoa	**8** Mt St Helens	**15** Hekla
2 Taal	**9** Paracutín	**16** Eldfell
3 Mayon	**10** Cotopaxi	**17** Beerenberg
4 Fuji	**11** Pelée	**18** Tristan da Cunha
5 Mauna Loa	**12** Villarrica	**19** Vesuvius
6 Kilauea	**13** Fayal	**20** Stromboli
7 Katami	**14** Teide	**21** Etna **22** Thira

ABOVE Distribution of the world's active volcanoes. The strong concentration along continental margins and mid-oceanic ridges is clearly shown.

ABOVE LEFT Active volcanicity at White Mountain, New Zealand.

ABOVE RIGHT Recent ropy 'pahoehoe' lava, Hawaii.

ABOVE Recent blocky or 'aa' lava, Meru volcano, Tanzania.

hand, shows signs of active volcanism and fracturing; Mars may have been active until quite recently and indeed may still be so. Earth, as we know, is still very active: a fact most strongly emphasized by the effects of earthquakes and volcanoes. In recent decades, however, geophysicists have learned of more subtle movements of the Earth's outer layers and the mobile mantle beneath.

Whenever molten matter (magma) rises toward or indeed reaches the Earth's surface, it carries with it dissolved gases. In moving toward the surface, it feels the effects of less and less load pressure as the thickness of overlying rocks decreases. Eventually a point is reached when the pressure exerted by the dissolved gases exceeds that of the rock burden, whereupon the gases come out of solution and form bubbles. The formation of bubbles, termed vesiculation, is one of a number of phenomena that help magma to rise to the surface, and may also have a marked effect on the volcanic landforms that develop.

If we study a map showing the distribution of both active and recently active volcanoes, we can notice a number of interesting features. Firstly, a very large number are actually located in the oceanic regions. In these locations some volcanicity may be concentrated along vast submarine mountain ranges, called 'oceanic ridges', but the volcanoes tend to be found in linear arcs close to or actually within the margins of continents: along the western edge of the Americas, for instance, or as arcs of oceanic islands such as are found in Indonesia and Japan. Then there is a smaller number that occur in association with rift valleys, such as the East African Rift.

Volcanicity – 2

Each of the terrestrial planets and at least one satellite of Jupiter shows the effects of volcanism. Of these the Earth and Io, Jupiter's moon, are currently active and there is strong evidence to suggest that this may also be true for Venus. The volcanic landforms encountered are very diverse, and so it is appropriate that we enquire what factors govern their location and morphology, since these may provide us with clues about the internal workings of the Earth and other planets.

The form of a volcanic structure depends predominantly on the chemical composition of the magma produced during an eruption and on exactly how this is released at the surface. Lavas containing low amounts of silica, like basalts, are generally quite fluid (non-viscous), while silica-rich lavas, such as dacites and rhyolites, tend to be very sticky (highly viscous). The former flow easily and may spread over wide areas; the latter generally give rise to thick stubby flows which may be explosively disrupted before they even reach the surface.

Viscosity, linked closely with magma's composition, is of fundamental importance in determining volcanic form. Resistance to flow – which is what viscosity is essentially – is also a function of the proportion of volatiles a lava contains and of its temperature. When vesiculation occurs, gas bubbles are formed in the molten lava. As soon as this happens the viscosity of the lava can increase quite dramatically, and many highly volatile lavas erupt explosively, generating explosion craters.

In time the lava will cool. Heat is lost both to the atmosphere and to the cold crustal rocks through which it is rising or over which it is flowing. As the temperature falls, so the viscosity increases and eventually a point is reached when it begins to solidify. This stage is marked by the rapid growth of crystals in the magma. When all of the liquid has been converted to crystals, an igneous rock has been created.

Where eruption is confined to a single vent or small group of vents over a long period, large volcanic structures are formed. Thus the Hawaiian islands consist of five huge basaltic volcanoes that have been active for several million years. Africa's highest mountain, Kilimanjaro, is another complex structure where volcanic activity has been localized over a lengthy period.

The most abundant volcanic rock on Earth – and almost certainly on the other terrestrial planets – is basalt. Basaltic lavas contain relatively little silica but a lot of iron compared with other types. They are characteristically very fluid and often issue from long fissures, giving rise to widespread flat topography, as is seen in the Columbia River region of the United States and in the Deccan Traps of India. Where they are erupted from localized vents, as in Hawaii, they tend to form low-profile volcanoes called shields. These may be very large, and where several occur in a relatively small region they may either overlap or coalesce.

Lavas with higher silica content, being more viscous, build steeper-sided volcanoes, like Kilimanjaro and its close neighbour Meru to the west. Mount St Helens and Lassen Peak, volcanoes in the Cascades Range of the western United States, fall into a similar category, although their lavas are somewhat different from those of their African counterparts. Eruptions involving these more siliceous magmas are often explosive and may be devastating: witness the 1980 event on Mount St Helens!

The classical conical volcanic form is typical of a significant percentage of volcanoes, but huge volumes of lavas are erupted from fissures and never produce cones. This is true of very many of the submarine centres, particularly those situated on mid-oceanic ridges, and it is a style of eruption typical of vast areas on our neighbouring planets. These basaltic plains and plateaux are constructed by scores, perhaps hundreds, of eruptions that issue from cracks, fissures and small vents, and which over millions of years contribute to the construction of planetary crusts.

6

MAGMA

What is Magma?

Most people will have seen pictures of molten lava escaping from a volcano or volcanic fissure. This is the surface expression of volcanism, a process which is instigated within a planet by a process called partial melting. Let us begin by answering the simple question, 'What is magma?'

Magma is molten rock. It is the material from which all igneous rocks crystallize. Ignoring the details of exactly how magmas originate for the moment, a body of magma will be at high temperature and under considerable pressure when it is within the Earth. As it moves toward the surface, however, it will become cooler, losing heat by conduction to the rocks it passes through, until eventually it will cool sufficiently for crystals to form within it. Before this point is reached there is no regular arrangement of the chemical elements it contains, the various atoms moving around fairly freely. As soon as the freezing temperature of the more refractory minerals is attained, however, the atoms become very sluggish and eventually will be unable to move at all.

Terrestrial magmas nearly always contain silica, that is, silicon combined with oxygen. Then there are varying amounts of what are called 'major elements', such as aluminium, titanium, iron, magnesium, calcium, sodium, potassium, phosphorous and manganese. Generally speaking silica accounts for between 35 and 75 per cent of the whole depending on the type of magma; basalts usually contain between around 40 and 48 per cent silica, while granitic magmas have between 68 and 75 per cent. In addition to these elements, whose concentrations are usually measured in per cent of the total weight, there are much smaller amounts of 'trace elements', such as nickel, cobalt and rubidium, which occur in amounts normally measurable only in parts per million. Then there is a significant volume of volatiles, the most important of which is always water, accompanied by such elements as boron, chlorine, fluorine and compounds containing sulphur, hydrogen, oxygen and nitrogen.

While the magma stays deep within the Earth, these mobile elements remain dissolved in the magma, but as it rises toward the surface and the hydrostatic pressure becomes less and less, the more volatile components of the magma may be released as gases, like carbon dioxide and hydrogen sulphide. Thus when a magma actually reaches the Earth's surface it generally consists of a mixture of molten lava and solids and gases.

Inside the Earth

The discovery of the major internal divisions of the Earth was made through study of earthquake waves. These are generated when the rocks slip suddenly with respect to one another, usually along fractures called faults. Three kinds of wave are produced: compressional

The rise of magma. As hot fluid magma rises it cools. Crystals form as it nears the surface and the contained gases exsolve from it, forming bubbles. This bubbling, or vesiculation, aids its rise to the surface.

Rising magma

waves (P-waves) which move in a push-pull manner; shear waves (S-waves) which vibrate back and forth in a direction perpendicular to their travel path; and longitudinal waves (L-waves) which travel along the Earth's surface. Both P- and S-waves can be transmitted freely through solid rocks but only the S-waves can pass through liquids. This has been found a useful property when considering the nature of the Earth's interior.

Earthquake or 'seismic' waves behave rather like light waves in that they can be reflected from, transmitted through or refracted within layers of rock inside the Earth. The denser the material, the greater the velocity with which the waves travel; when they encounter a change in density of the medium through which they pass, they will be refracted, just as rays of light are when they pass through a lens. By analysing the effects the internal rocks have upon seismic waves, geophysicists have built a picture of the interior of the Earth.

Natural seismic waves take a variety of different paths through the rocks. The foci of earthquakes are at varying depths; the paths which the generated waves take, and the time they require to reach the surface, can be determined using sensitive instruments called seismographs. A high degree of accuracy is achieved by setting up seismographic stations all over the Earth's surface, and recording how long

Inside the Earth. Diagram showing the internal layering and the way in which temperature, pressure and density increase with depth. Sudden changes are accompanied by abrupt alterations in the velocity of seismic waves at 'seismic discontinuities'. These separate our planet into a number of discrete layers.

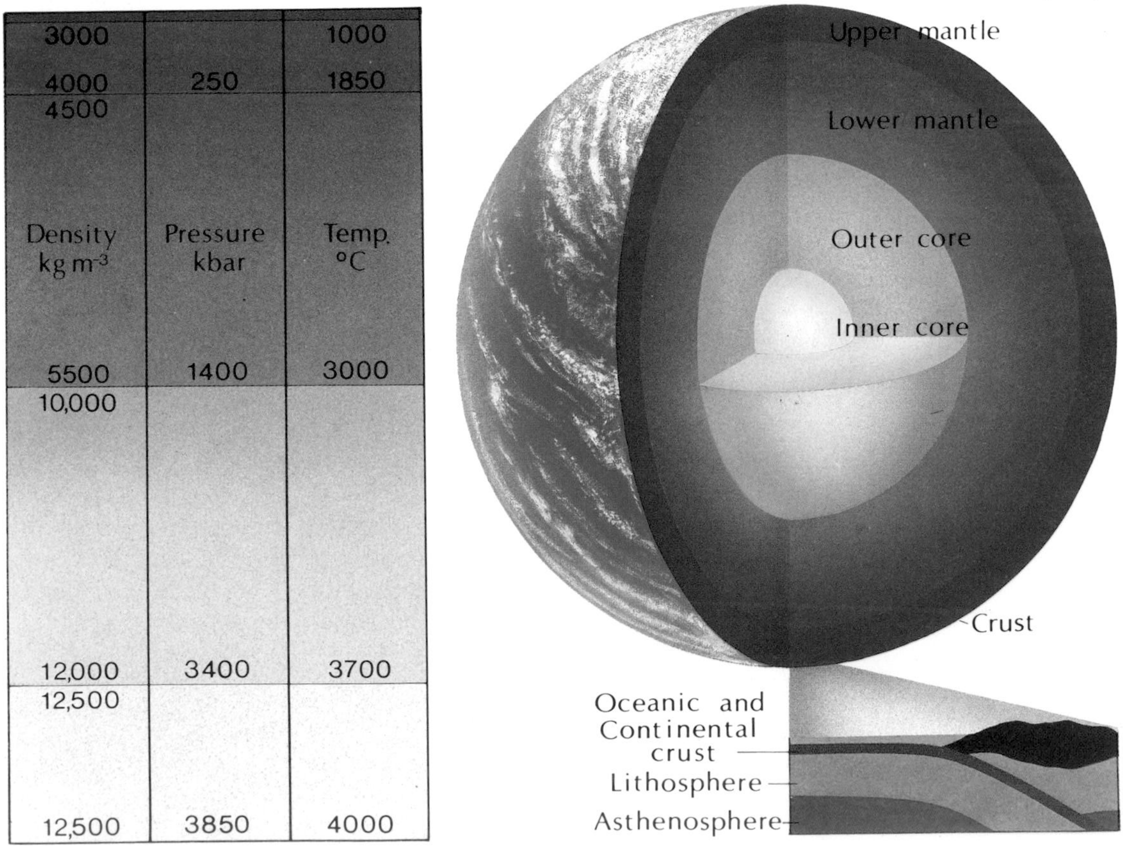

Density kg m^{-3}	Pressure kbar	Temp. °C
3000		1000
4000	250	1850
4500		
5500	1400	3000
10,000		
12,000	3400	3700
12,500		
12,500	3850	4000

Upper mantle

Lower mantle

Outer core

Inner core

Crust

Oceanic and Continental crust

Lithosphere

Asthenosphere

the waves generated by a quake take to reach the different recording stations. In this way, the exact location of the event can be established.

The outer boundary of the Earth's core strongly reflects both P- and S-waves; in fact the S-waves cannot pass through it. This is because the outer part is liquid and cannot transmit these kinds of waves. In fact, major seismic 'discontinuities' exist wherever there is an obvious change in either rock type or the properties of the rocks encountered.

It has been discovered that the mantle can be subdivided into a whole series of layers, whose individual boundaries could be established because at their boundaries, the velocity of seismic waves changes abruptly. In particular the outermost 70 km of the Earth could be separated from the underlying layers; as seismic waves passed below this level, their velocity altered sharply. This boundary defines the base of the lithosphere. The oceanic and continental crust we can directly sample lies above the main part of the lithosphere.

The characteristics of the lithosphere are strength and solidity, but the underlying zone is a relatively weak one and seismic waves are strongly attenuated as they pass through; this is why the layer became known as the asthenosphere or 'Low Velocity Zone'. The base of the asthenosphere is at a depth of 400 km. It seems to be a stiff mixture of crystals and magma; chemically it is very similar to the lithosphere, which is probably made of the rock peridotite. With increasing depth, the velocity of seismic waves increases again, the rocks becoming more solid then, at a depth of about 400 km, there is a sudden increase in density. This puzzled geologists for some time, but an experiment involving study of the mineral olivine under very high pressures confirmed the idea that here we have what is termed a 'phase change', where material of the same composition becomes so compressed that its atoms pack together more closely. In the experimental work, olivine changed to spinel and it is probably this phase transition that causes the velocity change inside the Earth at this depth.

Beneath this phase change horizon there is another quite thick layer which shows a gradual increase in density with depth; then another transitional zone below which density again increases until the core-mantle boundary, at a depth of about 2900 km.

The Origin of Magmas

Magmas exist, of that there is no doubt, but exactly where and how do they form? Perhaps the first question to answer is where they are currently emerging at the Earth's surface. The most active sites, yet least accessible to human view, are what are called oceanic ridges: linear, predominantly submarine mountains that run across the floors of the oceans. Rifts, fissures and vents situated along these ridges extrude considerable volumes of basalt. Indeed the bulk of modern basalts appear to be emerging at these sites.

Basalts are also found on the continents, but typically the more siliceous magma types occur here. Clearly there must be underlying reasons for this fact.

When they crystallize, magmas consist of a variety of silicate minerals, each of which has its own specific melting point. If a magma contains substantial amounts of water, however, the mineral melting points are substantially lower than they would be if the magma were 'drier'. The presence of water, therefore, allows rocks to melt at lower temperatures than they might otherwise do.

Commonly associated with magmatic bodies that solidify below the surface are veins which radiate away from them. These are typically infilled with what are called 'hydrated' minerals, that is, minerals that bind water into their atomic lattice. Some of this water undoubtedly originated within the magma itself but a lot may also have been contributed from the adjacent rocks, in which groundwater was circulating at the same time as the magma was being emplaced.

Temperatures and pressures in the upper regions of the lithosphere are generally too low for rocks to be melted. At greater depths, however – say around 70-100 km – we might expect at least some melting even of 'dry' rocks. At such depths we would be dealing with mantle material under considerable pressure and at elevated temperatures. The predicted melting of rocks at such depths appears to be borne out by seismic data that indicates a partially molten zone in the asthenosphere at these depths. Geologists believe that it is here that most basaltic magmas are generated, probably by selective melting of some components of the mantle layer.

The hydrated silica-rich magmas common in many parts of the Earth's crust, particularly the continental regions, cannot have been formed by simple melting of dry mantle material. Modern ideas generally appeal to the notion that they are produced by the partial melting of a variety of water-rich crustal rocks, like quartz sandstones or even existing igneous rocks. Some kinds of magmas, particularly those of andesitic composition (moderate silica content), which are very typical of active tectonic belts such as those that encircle the Pacific, may have their origins in a combination of partial melting of mantle rocks and partial melting of sediments or igneous rocks rich in silica and alumina. In the presence of substantial amounts of water, such a mix could probably produce silica-rich magmas at depths of as little as 35 to 40 km. As we shall see in a later section, the ideal sites for such a process are encountered at what are called subduction zones.

The Lesson of Hawaii

Over a century ago the famous American geologist J. D. Dana noted that there was an increase in age along the Hawaiian chain of islands, from the active volcano of Hawaii in the southeast, to the heavily eroded island of Niihau to the northwest. A similar trend was noted for the Canary Islands in the Atlantic and has subsequently been suggested for several other of the Pacific chains. The Hawaiian volcanoes are undoubtedly the most closely studied on the Earth and so it is logical to enquire what has been learned from all this investigation.

The Hawaiian islands form the southeastern end of a long series of atolls, reefs, submarine mountains and submerged volcanic cones that extends for 6500 km across the Pacific Ocean. The volcanoes within this chain run stepwise rather than in a simple line across the ocean floor, the individual eruptive centres being spaced roughly 75 km apart. Most of the islands are built from two or more volcanoes, Hawaii itself being formed from five: Kilauea, Mauna Loa, Mauna Kea, Hualalai and Kohala. The greater volume of each of the volcanoes is submerged beneath the ocean. Thus the oceanic rise upon which the chain sits is elevated about 6000 m above the oceanic floor, and then a further 4000 m above sea level on the island of Hawaii itself. The flanks of the volcanoes have gentle slopes, ranging between 5 and 10 degrees, the steepness tending to increase below sea level.

It is the age pattern shown by the islands that is of extreme interest: there is a gradual increase in age from Hawaii in the southeast, where

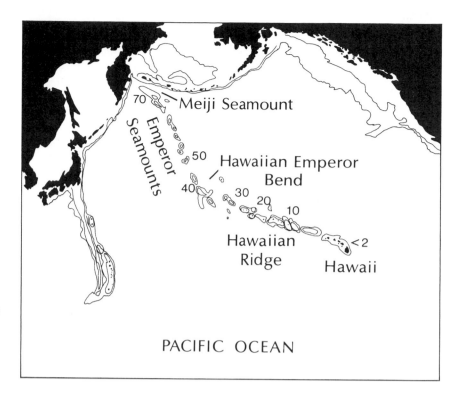

PACIFIC OCEAN

Ages of the volcanoes in the Hawaiian-Emperor chain (millions of years). Currently volcanoes on the island of Hawaii itself are active; with increasing distance towards the northwest end of the chain, however, successively older centres are encountered. Some of these have been extinct for over 30 million years.

there is current activity associated with the centres of Kilauea and Mauna Loa, until we reach the 5.8 million-year-old centre of Kauai, some 520 km northwest. When age data are considered for more distant members of the chain, it is clear that activity has been going on for at least 65 million years.

What appears to have happened during this long period is that the focus of volcanic activity has apparently slowly shifted southeastward, at a rate calculated to be around 8 cm per year. In 1963, J.Tuzo Wilson expounded his famous theory that the age variation of the Hawaiian chain developed as a result of the lithosphere in this region moving slowly but continually northwestward over a fixed 'hot spot' in the mantle beneath. He suggested that under the island of Hawaii was a currently active upwelling of hot mantle material that has been in existence for many millions of years and has supplied magma to the surface more or less continually during this time. Volcanoes that formed over the magma source were gradually moved northwestward away from the source, eventually becoming extinct as the magma supply was cut off from them.

The hot spot idea has become a widely accepted hypothesis and modern estimates suggest that the Hawaiian one has a diameter of about 300 km and is situated some 50 km northeast of Kilauea volcano. Other workers have extended Wilson's ideas, suggesting that hot spots are the result of upward movement of a 'plume' of hot mantle material from great depth within the mantle, this being driven by upward convection: the 'heat engine' at work again! Typically, hot spots express themselves at the Earth's surface in widespread volcanism, positive gravity anomalies, a higher than normal outflow of heat from the interior, while the mantle beneath has unusual characteristics.

Alternative explanations have been put forward. Most, however, are less readily supportable than the mantle plume idea described. Certainly the explanation we cite here can be nicely accommodated into the theory of plate tectonics.

Minerals and Magmas – 1

The two elements, oxygen and silicon, are by far the most abundant in the terrestrial crust and mantle. Thus it is that when a magma crystallizes, the majority of the minerals formed are silicates, that is, combinations of silicon, oxygen and other metallic elements. These rock-forming silicates show a wide range of chemical composition, external appearance and physical properties, but each is based on an internal structure called the silicate tetrahedron.

Silicon and oxygen combine naturally in the ratio of one silicon atom to four oxygens (SiO_4), the latter being positioned around the solitary silicon atom so that they form the four apices of a three-dimensional figure called a tetrahedron. This SiO_4 'anion' is extremely versatile since it can form complex chains and sheets of tetrahedra, forming giant molecules. In these, some or all of the oxygens belonging to individual tetrahedra are shared with adjacent ones; the precise way in which they are arranged gives rise to a small number of silicate mineral 'families'. Within these families the chains, rings or sheets of silicate anions are linked together by a variety of metal 'cations', like iron and calcium.

In the simpler families, the individual tetrahedra remain rather as islands, sharing none of their oxygens and simply being joined together by cations like iron and magnesium. With the sharing of more and more of the oxygens between adjacent tetrahedra, larger holes develop in the crystal lattice and larger cations, like calcium, sodium and potassium, can be accommodated. Indeed, in silicates belonging to the amphibole and mica families, even the large (OH) hydroxyl ion can be fitted. These kinds of silicates are therefore less dense than the simpler more compact ones such as olivine and pyroxene.

Because the electrostatic bonding between the silicon and oxygen atoms of the silicate tetrahedra is substantially stronger than that between the tetrahedra and the other metallic cations, crystal morphology, and lines of weakness (cleavages) within the crystals of the various silicate minerals, are predominantly a function of the SiO_4 tetrahedral arrangement. The various crystal frameworks are shown in the accompanying diagrams.

Minerals and Magmas – 2

The silicate minerals have their origins in the crystallization of magmas; thus, as a magma cools, crystals begin to form. The order in which specific minerals appear depends on the temperature and pressure, and on the amount of volatiles the magma contains or is able to absorb from surrounding water-rich rocks. Generally those minerals with the simpler structures will crystallize first; later, as the magma cools, more complex silicates emerge.

Not all of the minerals associated with magmas are silicates. Within most bodies of the igneous rock there are small amounts of non-silicates, like oxides, carbonates and sulphides. Exceptionally some oxides may accumulate as layers of crystals within a magma body, producing important ore deposits, like the chromite deposits of the Bushveld intrusion of South Africa. Most of these non-silicates, however, appear to be carried away from their parent magmas by volatile-rich fluids which may 'mineralize' the surrounding rocks.

Most of the metals such as iron, magnesium and zinc that may eventually be concentrated as ore deposits – which we may define as economically workable sources of metalliferous materials – appear to

TOP Purple fluorite crystals embedded in whitish calcite.

CENTRE Black tourmaline crystals set in white plagioclase and quartz.

ABOVE Crystals of the iron oxide, haematite.

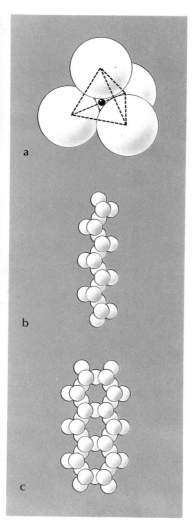

Silicate atomic structures:
(a) basic silica tetrahedron of
four oxygen atoms surrounding
one silicon atom; (b) single
chain silicate; (c) double chain
structure.

leave their parent magmas as metal chlorides. These are held in solution in the hot aqueous fluids. These fluids will be very hot as they leave the magma body and travel toward the Earth's surface along fractures or through permeable rocks. As they travel through the cooler crustal rocks, so their temperature falls and they enter into chemical reactions of various kinds with the rocks through which they pass. At successive stages in their upward passage, different metalliferous minerals are deposited, usually in order of decreasing melting point.

The deposition of ore minerals, and of the numerous non-metallic minerals - called gangue minerals – which usually accompany them, is a very lengthy process by human standards. For a small ore body, deposition will take decades, while a large one will need millenia to form. During this time the metal-bearing fluids cool continually, albeit slowly, so that eventually, even near to the magma source, temperatures will have dropped many hundreds of degrees. This being so, the high-temperature minerals precipitated near the magma source during the early stages of mineralization will probably be joined by lower-temperature minerals crystallizing much later on, giving rise to what is called a zoned ore body.

The study of such bodies is a difficult one, but because we urgently need metals for our everyday existence, they have been intensively researched. We know that the actual processes operating within the Earth are very complex, and that we do not know the answers to all of the questions posed by ore deposition.

How Hot is it Inside the Earth?

We have been talking glibly about high temperatures inside the Earth and how minerals and magmas form, but as yet we have avoided asking the question, 'How hot really is it inside our planet?' Measurements made with specially adapted thermometers in deep mines and bore holes indicate that the average rise in temperature per 100 m of depth is between 2° and 3°C. Temperature gradients can only be measured in this way down to depths of around 8 km, however, beyond which other methods have to be utilized.

Geophysicists have to use sophisticated means to discover exactly how much heat is flowing out from the Earth, but in recent years they have been able to measure, at least for the continental crust, what the surface heat flow is. Having done this, calculations can be made which reveal that in tectonically active parts of the continental crust, temperatures of the order of 1000°C are encountered at depths of around 40 km. In contrast, at the same depth under more stable parts of the continents, the geothermal gradient is less and the temperature rises only to around 500°C. Significantly, volcanism and crustal mobility are associated with areas of high heat flow, whereas relative stability is typical of those regions of lower heat flow. Under the oceans, the outflow of heat is greatest along the oceanic ridges where seismic activity and volcanism are concentrated. Elsewhere the temperature on the ocean floor is probably close to 0°C, rising to about 1200°C at the base of the lithosphere.

At present there is no direct way of telling what the temperatures are at greater depths, but the indications are that at the Earth's centre the temperature approaches 4300°C, while at the core/mantle boundary a figure of approximately 3700°C is likely. To find a way of making accurate determinations at these great depths is just one of the as yet unsolved tasks for geologists.

7

THE PERPETUAL DYNAMO

Magnetism

Lodestone, with its power to attract or repel pieces of iron, has been known since ancient times. It is in fact an oxide of iron called magnetite. The Chinese seem to have known 3000 years ago that a magnet will point toward the North Pole when freely suspended and allowed to swing in any direction. Yet this knowledge escaped the Western world until the twelfth century, when it was first used in that invaluable navigational aid, the mariner's compass.

In the earliest form, the water compass, a piece of magnetized iron was placed in a wooden vessel and floated on water, so that it could move freely in a horizontal direction. Later mariners used a pivoted magnetic needle with a disk surmounting it. This was known as a compass card, and was divided into 32 equal 'points'.

In 1269 Picard Petrus Peregrinus described how he located the magnetic poles of a lodestone globe, and also showed how like poles repel one another while unlike poles attract. Famous mariners such as Columbus and Magellan used compasses, but the first really major advance was made by the English physicist William Gilbert. His book *De Magnete, Magneticisque Corporibus*, published in 1600, discussed magnetic variation and contained the suggestion that the Earth itself behaves in the manner of a huge magnet.

Gilbert could not explain how a compass worked, but in the early nineteenth century experiments with simple batteries showed how

A late-Victorian painting by Arthur Ackland Hunt showing the distinguished physician William Gilbert, demonstrating some of his experiments into magnetism at the court of Queen Elizabeth I. Gilbert was one of the first scientists to realize the Earth behaved like a huge magnet.

magnetic forces are generated. Magnetism was, in fact, a function of electrical forces. These should really be termed *electrodynamic* forces, but for historical reasons they are normally known as magnetic forces.

The Earth's Magnetic Field

As Gilbert had suggested, the Earth behaves like a huge magnet; if a magnetized needle is suspended freely, it will align itself in the direction of the local magnetic field. The easiest way to represent a magnetic field is by lines of force; that is to say, by a series of lines everywhere pointing in the direction of the field. Where the field is strong, the lines of force will be crowded together; with a weak field the lines are more widely separated. The lines are, therefore, contours of the strength of the field.

The Earth's field takes the form of a magnetic dipole, behaving in the same way as the field produced by a bar magnet. Indeed, the best way to understand the features of the Earth's field is to picture a huge bar magnet running right through the centre of the Earth itself. To be more specific, the axis of the magnet is almost parallel to the geographical north-south line or spin axis; at present the angular difference is about 18 degrees. The north magnetic pole is located in the Arctic islands of Canada. The south magnetic pole lies south of Tasmania, although, as will be seen in a later section, this has not always been so; the positions of the poles are not constant.

The field strength at any point on the Earth's surface is measured by a delicate instrument known as a magnetometer. The unit of field strength is the *gauss*, named after the famous nineteenth-century German mathematician J. K. F. Gauss. At the Equator, the field strength is about 0.3 gauss; at the magnetic poles it is roughly double this. To put this in perspective, it is worth noting that a typical horseshoe magnet has a field strength of about 10 gauss.

Although the picture of a huge bar magnet inside the Earth is a good analogy, it is an over-simplification, and it cannot be said that as yet we fully understand the real situation. However, this analogy is adequate for our present purpose.

The Earth's Magnetosphere

The magnetic lines of force act as a shield round our planet, protecting us from harmful radiations from space. Moreover, high-energy, high-velocity particles constantly strike the upper atmosphere. There are, in particular, cosmic rays, which are not really rays at all but are atomic nuclei; some of them come from the Sun and from the planet Jupiter, although most originate far beyond the solar system. When the cosmic ray particles strike the upper air, the heavy nuclei are broken up and only the harmless 'secondary' particles reach the ground.

To check the cosmic ray counts at various altitudes, the American artificial satellite Explorer 1 was launched in 1958. This was in fact the first successful American satellite, and proved to be more important scientifically than the early Russian Sputniks. Explorer 1 entered an elliptical orbit with a perigee (minimum distance from Earth) at 350 km, and its Geiger counter measured the cosmic ray intensities at different heights. The results were surprising. Each time Explorer 1 reached a height of 950 km, the readings dropped to zero. It was then found that this was due to radiation so intense that the Geiger counter had become saturated and put out of action. Later investigations showed that there are zones of intense radiation surrounding the

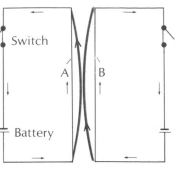

ABOVE LEFT Italian mariner's compass, 1580.

ABOVE RIGHT An illustration showing the effect of connecting simple batteries in different ways. In the first example (ABOVE) where the connections are as indicated by thin arrows, the strips of wire that complete the circuit are toward one another. If these connections are reversed (BELOW), the force between the wires becomes one of repulsion.

Earth; these are now known as the Van Allen radiation belts, after the American scientist James Van Allen, who had been responsible for the equipment on Explorer 1.

The Van Allen belts are contained in the magnetosphere, which is really an extension of the magnetic force field into a doughnut-shaped region surrounding the Earth. On the Sun-facing side it extends outward for about 10 Earth radii, but on the side away from the Sun it streams out into a 'tail' for several millions of kilometres before losing its identity in the overall magnetic field that pervades the solar system.

The asymmetry is due to the solar wind, a stream of ionized particles (plasma) pouring out from the Sun in all directions. Upon striking the Earth's force field the particles are decelerated, producing a 'bow shock' inside which there is a turbulent region known as the *magnetosheath*, where the pressures from the solar wind and the Earth's field are equal. Inside this again is an interface known as the *magnetopause*, which encases the magnetosphere proper.

Changes in the Magnetic Field

A magnetized needle taken to various points on the Earth's surface and allowed to swing freely will settle down at quite different angles to the horizontal. The dip shown by the needle is termed magnetic inclination. Moreover, the needle aligns itself not with the geographical pole,

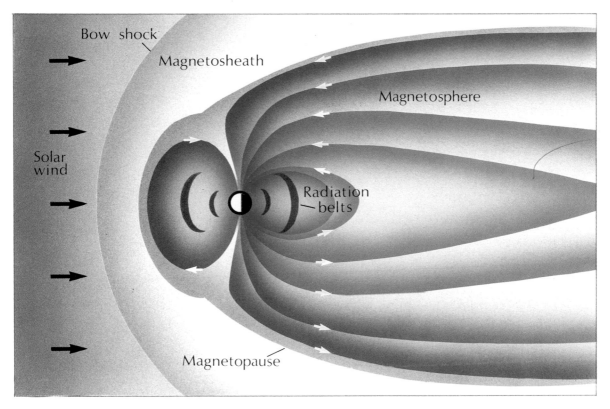

ABOVE The magnetosphere is roughly teardrop-shaped, with its tail extending far away from the Sun. The asymmetry is created by interaction between the Earth's magnetic field and the high-energy particles of the solar wind which stream out from the Sun. Where the two meet a shock wave is produced (bow shock) within which is a turbulent zone bounded on the inner side by the magnetopause. The magnetosphere proper lies on the Earthward side of the latter.

BELOW LEFT The Earth's magnetic field is best considered as being produced by a huge bar magnet at its centre. The alignment of the magnet does not quite coincide with that of the Earth's spin axis. A suspended magnetized needle would align itself in the direction of the magnetic field as shown.

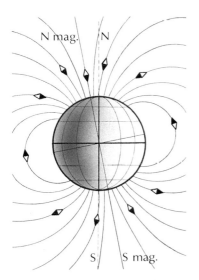

but with the magnetic pole. The angular difference is termed magnetic declination. As early as the sixteenth century it was found that the position of the magnetic pole changes with time; this is termed secular variation. Thus at the end of the fifteenth century the magnetic variation was 11.25 degrees, but fell to only 6 degrees by the end of the sixteenth century. Fairly good records exist over the past 400 years. If we retain the picture of a huge bar magnet through the Earth, this magnet must 'wobble'. Only a liquid core could move fast enough to produce changes of this order, and it is now generally accepted that magnetic forces do indeed originate in the Earth's core.

Rocks, Minerals and Magnetism

Iron is an essential ingredient of all natural magnets, but not all magnetic substances behave in the same way. Thus metallic iron (Fe) is said to be ferromagnetic, because all its tiny atomic magnets are aligned in the same direction, while minerals such as magnetite (Fe_3O_4) are ferrimagnetic, because some of their atomic magnets point

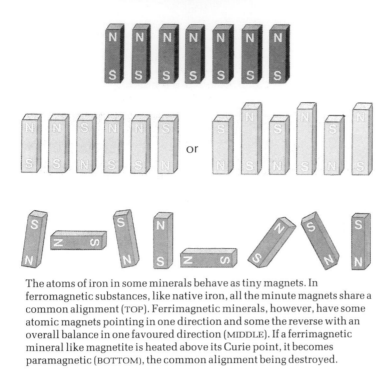

The atoms of iron in some minerals behave as tiny magnets. In ferromagnetic substances, like native iron, all the minute magnets share a common alignment (TOP). Ferrimagnetic minerals, however, have some atomic magnets pointing in one direction and some the reverse with an overall balance in one favoured direction (MIDDLE). If a ferrimagnetic mineral like magnetite is heated above its Curie point, it becomes paramagnetic (BOTTOM), the common alignment being destroyed.

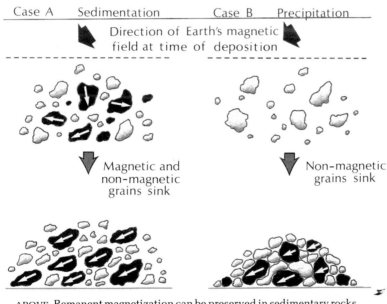

ABOVE Remanent magnetization can be preserved in sedimentary rocks. In case A (LEFT) magnetic grains settling through water will be deposited on the sea floor such that their internal magnets become aligned in the direction of the prevailing geomagnetic field. In case B (RIGHT) magnetic material is being precipitated in the pore spaces between detrital grains and also takes on the prevalent field direction.

in opposite directions. If a ferrimagnetic mineral is heated above a certain well-defined temperature its magnetization will be destroyed, and it becomes what is termed paramagnetic, with all the atomic magnets randomly orientated. This critical temperature is the Curie point, usually around 500°C.

Exactly how a rock becomes magnetized depends upon the way in which it was formed. Rapid cooling, as with hot lava extruded on to the

Earth's surface, leads to crystallization; the temperature will fall below the Curie point, and the atomic magnets will align themselves in the direction of the geomagnetic field (thermoremanent magnetization, or TRM). Other rocks may never become so hot, and will form by the settling of crystal fragments through water to form sedimentary rocks; if there are some already-magnetized grains in the deposit, they will tend to become aligned in the direction of the field. Magnetic minerals growing as chemical precipitates in sedimentary deposits will tend to grow with their magnetism aligned with that of the Earth (detrital remanent magnetization, or DRM). The Earth cannot be magnetized to great depths, because the temperature will be above the normal Curie point – which again destroys the simple picture of a bar magnet passing right through the globe.

Palaeomagnetism

Palaeomagnetism is the study of the fossil magnetization of rocks of all ages. It has been found that rocks containing magnetic materials will bear the imprint of ancient magnetism, and measurements have shown that the magnetic poles have not always been in their present position: a phenomenon termed polar wandering. It has been deduced from this fossil magnetism that not only have the continents moved relative to the magnetic poles but that the continents have moved relative to each other. Another important discovery is that the Earth's field has not always had the same polarity; at various times in the past a compass needle would have pointed south, not north. There have apparently been at least nine such reversals during the past 4,000,000 years.

Magnetometers carried on board ships or aircraft have detected magnetic anomalies beneath the ocean floor. A positive anomaly indicates that the rocks have the same polarity as the present geomagnetic

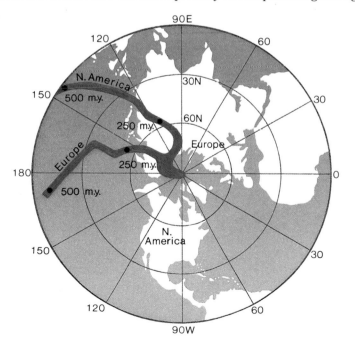

Map to show the apparent polar wandering curves for North America and Europe from 550 m.y. ago until the present day. The apparent wandering is a reflection of movements within the Earth's lithosphere, since the actual magnetic poles deviate little from the geographic poles.

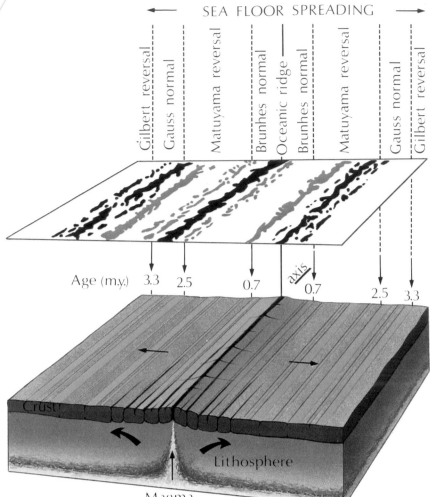

Gilbert reversal
Gauss normal
Matuyama reversal
Brunhes normal
Oceanic ridge
Brunhes normal
Matuyama reversal
Gauss normal
Gilbert reversal

Age (m.y.) 3.3 2.5 0.7 axis 0.7 2.5 3.3

Crust

Lithosphere

Magma

LEFT Polarity reversals are revealed by measuring magnetic anomalies in the basaltic rocks of the ocean floor. The alternating positive (green) and negative (brown) anomalies shown in the lower part of the diagram are related to linear belts of magnetized rocks on the sea floor which were produced at the site of an oceanic ridge (spreading axis), and then carried away from it on both flanks by convective movements which produced lateral motion of the lithospheric plates. Lavas extruded during periods when normal polarity prevailed give positive anomalies; those produced while the reverse applied show negative anomalies.

RIGHT A chart of the Atlantic sea floor showing how dating magnetic stripes reveals the symmetrical arrangement of the oceanic rocks about the spreading axis (Mid-Atlantic Ridge).

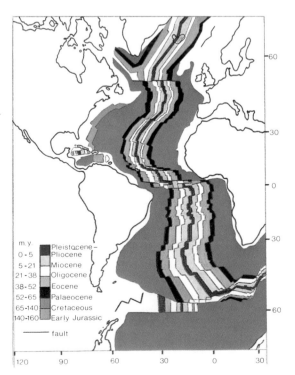

m.y.	
0 - 5	Pleistocene–Pliocene
5 - 21	Miocene
21 - 38	Oligocene
38 - 52	Eocene
52 - 65	Palaeocene
65 - 140	Cretaceous
140 - 160	Early Jurassic
——	fault

120 90 60 30 0 30

60 30 0 30 60

field; a negative anomaly indicates a reverse polarity. These anomalies alternate, and trace out magnetic stripes or linear patterns that extend for hundreds of kilometres over the sea floor.

We also know that submarine mountain chains (oceanic ridges) rise from the ocean depths. One of these, the Mid-Atlantic Ridge, divides the Atlantic along a north-south axis. To either side of this crest the magnetic stripes are almost perfectly symmetrical, and similar patterns are found for all the other known major oceanic ridges.

It seems that the basaltic rocks of the oceanic crust act like natural tape recorders, recording successive periods of normal and reversed polarity when the relevant rocks poured out on to the ocean floor. We may assume, therefore, that the oceanic ridge crests are linear lava conduits, through which hot basaltic magma rises and flows out on to the sea floor. The lava will cool, and its magnetization will be fossilized. Once fossilized, the polarity will not change; but the lava moves gradually away from the ridge axis as the sea floor spreads. Radiometric dating shows that the strips of basalt flanking an oceanic ridge are progressively older with increasing distance from the ridge crest.

We can therefore determine the times of magnetic reversals. For example, a basalt extruded from the Mid-Atlantic Ridge about 10.5 million years ago shows reversed polarity, so that the field of the Earth itself had reversed polarity at that time. This supports the theory of sea-floor spreading. For the Atlantic the present rate is about 1 cm per year in each direction.

The Perpetual Dynamo

The generally accepted theory of the Earth's magnetic field is due largely to the work of Sir Edward Bullard and W. M. Elsasser. It hinges primarily on the probability that the outer part of the Earth's core is composed of liquid metal: iron and nickel. This means that it can conduct electrical currents. If the fluid is in motion, it can also interact with a magnetic field, and the field itself can influence movements in its interior.

Weak magnetic fields exist in the solar system, and indeed all over the Galaxy. If for some reason or other the Earth's core were in motion, even a weak field of this kind could influence its movements. If the pattern of movement happened to be just right, then a magnetic field could be created within the core. It is now generally believed that the Earth had no magnetic field at all in its earliest period of separate existence, but that a field was subsequently developed by an interaction of this type.

The Earth has a fairly rapid axial rotation. This spin has a profound effect upon the fluid motions in the core, and results in the Earth behaving in the manner of a huge dynamo. Technically it is said to be a 'self-exciting dynamo', since once it begins to work it gathers momentum, regenerates the weak galactic field, and produces the relatively strong field of today. The exact mechanism is very complex, and is not yet fully understood; note, moreover, that there are suspicions that around the times of reversals of polarity the overall magnetic field of the Earth may become very weak, perhaps vanishing entirely for a time – which could have disastrous effects upon life, since the protective 'screen' would be temporarily withdrawn. However, the fact that the magnetic poles are not very far from the geographical poles lends strong support to the idea of a dynamo process which is sustained largely by the Earth's axial spin.

A simple diagram showing how an electric current is generated when a copper disk is rotated through the magnetic field produced by a bar magnet (a). If a coil is substituted for the bar magnet (b) the same electrical current produces a magnetic field which perpetuates the system. The terrestrial magnetic field is believed to operate in a similar, if more complex way.

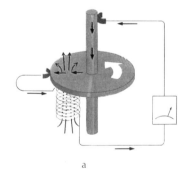

a

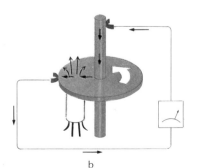

b

8

CONTINENTS
AND OCEANS

Oceanic Crust

The Earth's crust is not of the same thickness everywhere. Beneath the continents it varies between 10 and 75 km, but it is substantially less underneath the oceans, at most being about 8 km thick. There are other differences, too. For a start, the top of the oceanic crust is, on average, between 3 and 4 km lower than its continental counterpart. Then, there are seismic data which show that beneath the continents there are at least two major layers, the higher of which is of lower density. Underneath the waters of the oceans is what we call oceanic crust. It has a mean density of between 3.0 and 3.1 g cm^{-3} and is composed essentially of basaltic rocks.

The ocean floors are veneered with sediments, immediately beneath which are basaltic rocks forming a layer between 0.5 and 2.5 km thick. These in turn are underlain by about 5 km of coarser rocks called gabbros, which are separated from the mantle by a rather thin layer of olivine-bearing igneous rocks of somewhat greater density. Together these four thin layers comprise the oceanic crust.

Within the regions of oceanic crust there are major topographic features like the oceanic ridges, to which reference has already been made. Together these form an 80,000 km long chain of mountains rising from the ocean depths. Current volcanic activity is characteristic of these, and indeed it is here that new oceanic crust is currently being generated. The ridges rise between 2 and 3 km above the ocean floors, and often are offset by lines of fractures which may be traced for several thousand kilometres. Finally there are chains of islands, like the Hawaiian group, which are located well away from both ridges and fractures. These are also currently active, both volcanically and seismically, and are particularly characteristic of the Pacific Ocean.

Continental Crust

The crust beneath the continents is much thicker than that under the oceans and its surface is generally at much higher elevation. It is far easier to study its surface than that of the oceanic crust because it is not overlain by water. Expensive underwater operations are therefore not required to collect samples. The oldest piece of continental crust so far discovered is about 4000 million years old. This is in marked contrast to the oceanic crust, the oldest *in situ* slab of which is no older than about 200 million years.

Isostatic and gravity data both indicate that the upper part of the continental crust is composed of relatively less dense material, called sial after the two elements silicon and aluminium. Sialic rocks have a density between 2.7 and 2.8 g cm^{-3} and appear to be largely composed of silica-rich sedimentary rocks like sandstones, and metamorphic

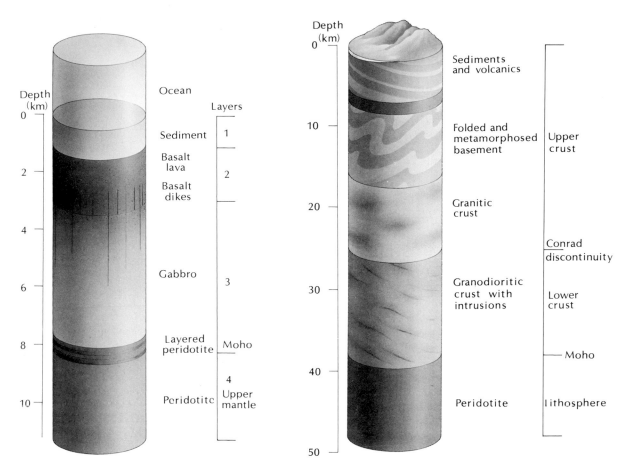

ABOVE LEFT A section through the Earth's oceanic crust, based on seismic data, dredged samples and interpretation of other geological data. The relatively thin layer of recent sediments is underlain by layers of denser basaltic rocks, which show a downward increase in density as they approach the mantle layer.

ABOVE RIGHT Section through the Earth's continental crust, based largely on seismic data. Compared with the oceanic crust, continental rocks are of lesser density and higher silica content.

rocks like gneiss and igneous rocks like granite. They are buoyed up on a layer of denser material which is similar to the oceanic crust. These two crustal layers, together with the upper part of the mantle, comprise the outer skin of the Earth called the lithosphere.

The actual boundary between the crust and mantle is defined by the Mohorovičić Discontinuity or 'Moho', a thin zone where earthquake waves experience an increase in velocity from 6.8 to 8.1 km s^{-1} in passing through; this corresponds to a marked change in composition. The crust thus comprises all of the rocks overlying the Moho, which means the sialic materials and the denser basaltic layer. Generally the continental crust is thickest under high mountains – as we should expect from isostasy; and it is thinnest close to the margins of the continents, where it grades into oceanic crust. The continental crust may also be extremely thin beneath large rift valleys under which the continental crust has been stretched.

Continents, Oceans and Gravity

If the Earth were covered uniformly by water, the surface of the water would define a surface known as the geoid. By observing the way in which the orbits of artificial satellites are disturbed by the Earth, it can be shown that this imaginary surface shows considerable departure from anything as simple as an ellipsoid. Map makers, who are particularly concerned about the exact form of the geoid when trying to establish the precise latitude and longitude of points on the Earth's surface, have settled for a calculated solution, known as the 'spheroid'. Broadly speaking, it corresponds to an average geoid.

Points that lie at the same latitude have a theoretical gravitational

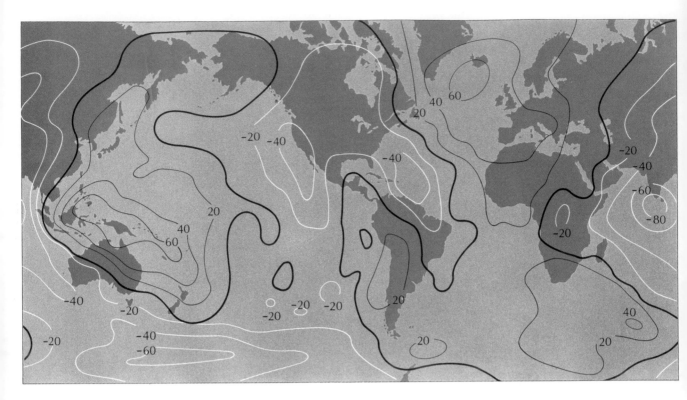

pull dependent solely upon latitude. By measuring the Earth's gravity with sensitive gravimeters, geophysicists can determine the actual gravitational force at any location and, after making a number of corrections, they can arrive at a figure that can be compared meaningfully with the theoretical gravitational value for the same point. This kind of work reveals for instance that the rocks beneath mountainous continental regions are less dense than those below lowlands and the oceanic areas. The disparities between the computed and measured values for gravity are called gravity anomalies.

When gravity anomalies are considered on a worldwide basis, it is found that the anomalies are generally positive over the oceans (there is an excess of gravity over the predicted value), while on the continents, and particularly over the sites of high mountains, anomalies are negative (deficiency of gravity). The higher the mountain, the higher the negative anomaly becomes. This confirms what has been said in the previous section concerning the nature of the oceanic and continental crust. By studying gravity data from thousands of stations at the Earth's surface, gravity maps have been drawn; these allow geologists to define regions where unusual values of gravity occur, and stimulate them into trying to explain what this means in terms of the overall workings of the Earth.

The Structure of the Oceanic Regions

Since the late 1930s a variety of new techniques has been developed, the result of which has been an explosive increase in our knowledge of the ocean floor. Gravity measurements made from orbiting satellites and, in particular, a new kind of aerial mapping, called geotectonic imagery, in which very accurate measurements of the height of the sea surface are made by a satellite-borne radar altimeter, have greatly refined our data. Furthermore new sonic mapping techniques have

World gravity map. Gravity varies from place to place on the Earth's surface. This is a reflection of the asymmetrical distribution of mass in the crust and mantle layers. The 'geoid' is the term for the shape which most closely approximates the ocean surface. Departures from this ideal form are indicated on the map by regions of positive and negative gravity.

RIGHT Section through the oceanic lithosphere, showing the rise of basaltic magma from the mantle toward a spreading centre at a mid-oceanic ridge. The basaltic magmas originate in partial melting of magnesium-rich mantle material (peridotite), and rise toward the surface, some solidifying in fissures as dikes, others spreading out to form new sea floor. As the lithospheric plates on either side of the spreading centre move away, so the basalts cool, becoming more dense, and settle lower into the oceanic layer.

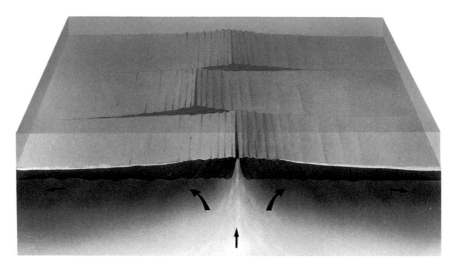

also added more detail to topographical charts of the ocean basins.

Modern theory says that the lithosphere is divided into a series of rigid 'plates' which move slowly over the more mobile asthenosphere below, rather as do wood rafts on a slowly moving river. The rate of relative movement of these plates is a few centimetres a year. Plate boundaries may be of three types: divergent, where adjacent slabs are moving away from each other; convergent, where they are moving

BELOW The major tectonic features of the Earth's crust.

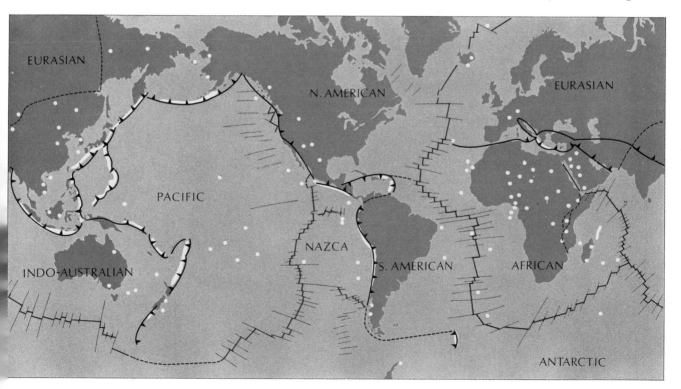

——— Constructive plate margin

——— Destructive plate margin

——— Undifferentiated margin

-------- Uncertain margin

——— Transform fault

——— Fracture zone

⚬ Hot spot

⬛ Oceanic trench

toward one another, one usually plunging down under the other at the boundary; and conservative, where adjacent lithospheric slabs slide past each other.

At divergent boundaries, such as those defined by oceanic ridges like the East Pacific Rise and the Mid-Atlantic Ridge, the lithosphere is being penetrated by basaltic magmas which have risen from the mantle, in response to the pulling apart of the lithosphere beneath the ridge axes. Rifts and fractures occur along the ridge crests and it is up such fractures that basaltic magma rises, flows out onto the flanks of the ridge, and eventually consolidates into new oceanic crust. While this is happening the sea floor is continually spreading away from the ridge axes, which is why they are known as spreading axes.

The complex strains imposed upon the lithosphere during plate motions induce much faulting of the oceanic crust adjacent to the oceanic ridges. Transform faults trend roughly normal to the ridge axis, while spreading faults open up parallel to the spreading axis itself. At points where transform faults and the spreading axis intersect, dome-like masses of lava appear to be forming, almost as if the new crust is being generated from numerous, fault-bounded feeders.

New Crust for Old

Beneath the spreading axes regions of the asthenosphere rise to within a few kilometres of the surface. This 'low velocity' mantle material has a peridotitic composition, being far richer in magnesium and iron and poorer in silica than typical crust. As this mantle material rises the pressure on the buried peridotite decreases, so that it expands upward, welling up from depths between 50 and 70 km to the base of the crust. During this deep-seated process there can be no significant loss of heat, so that as it rises the peridotite begins to melt. Less than a fifth of the volume of peridotite melts during this upward movement. The fraction that does melt has a basaltic composition since the temperatures reached will be close to the melting points of silicates like pyroxene, feldspar and olivine, and this accumulates in a reservoir somewhere near the base of the crust. By the process of fractional crystallization, whereby more refractory minerals separate from the magma and then sink toward the base of the reservoir, there is a concentration of denser silicates like olivine and calcium-rich feldspar to form layers of the plutonic rock, gabbro. It is these rocks that form the lower part of the oceanic crust.

Some of the magma, however, will rise upward, via a network of fractures, to the Earth's surface, there to be extruded onto the ocean floor as basaltic lava. A significant proportion of this basalt will also freeze in the fissures which either fed the flows or opened up at a later stage. As the new oceanic crust moves further away from ridge axes, it cools and contracts and its density increases. It will thus sink deeper below the surface of the ocean to maintain its isostatic balance with the underlying mantle. The depth to which this crust settles has been shown to vary with the square root of its age. Thus the older the crust, the lower it will have sunk, a fact that has proved useful in mapping crustal ages on the ocean floor.

Continental Structure

There is no oceanic crust older than about 200 million years old. This is because it is continually being destroyed and recycled at convergent plate boundaries. The continents, however, stand in strong contrast,

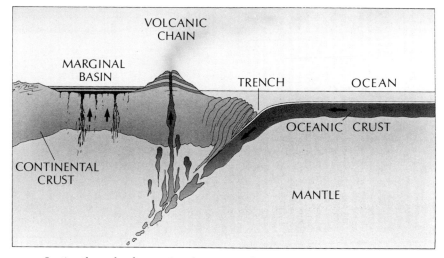

Section through a destructive plate margin showing the relationships between oceanic and continental crust. Where oceanic crust is being subducted beneath the margin of a continent, some of the downward-plunging slab melts, producing magma. Much of this cools into solid rock deep down but a proportion rises toward the surface and issues from volcanoes. Very often, behind the volcanic arc a marginal basin opens; this may subsequently be filled with sedimentary rocks and lavas produced during the plate movements.

since they generally resist this recycling and stand by imperiously, buoyed up on the denser substrate as slabs of oceanic lithosphere plunge beneath them. This destructive activity occurs along 'subduction zones'. At the type of subduction zone where a plate bearing continental crust at its leading edge converges upon a plate with only oceanic lithosphere, the denser oceanic material sinks down beneath the leading edge of the continental slab and is eventually destroyed by both mechanical forces and remelting as it plunges into the hot interior. Because of its resistance to subduction, a substantial volume of the continental crust is of greater age, the oldest rocks dating back nearly 4000 million years.

The rocks exposed on the continental regions can be divided into two main groups: (i) accumulations of sedimentary and volcanic rocks of wide aerial extent which show the effects of only minor deformation; (ii) strongly deformed belts of sedimentary, igneous and metamorphic rocks, called orogenic or mobile belts. The former do not occur everywhere but where they do, they always rest on the rocks of group (ii). In some places they are only a few kilometres in thickness.

The bulk of the continental crust has a complex history and has been affected by orogenesis at some time or another. Every orogenic belt represents a long series of events, spanning perhaps several hundred million years. Individually they represent vast chunks of geological history, which have been one of the main preoccupations of geologists. Belts which currently are adjacent, may differ in age by hundreds or even thousands of millions of years.

Many features of ancient orogenic belts closely resemble ones of much more recent age; in fact they have similarities with those evolving today. This gives us hope that careful study of modern rocks in such environments will eventually lead to a greater understanding of their much more ancient counterparts.

9

THE EARTH
FROM SPACE

Methods of Remote Sensing

As is so often the way with new areas of study, the first steps into the field of 'remote sensing', or indirect study of the Earth from the air, were made for strategic purposes in the mid-nineteenth century. Since those days, balloons, aircraft and orbiting satellites have been used for obtaining geological, oceanographical, meteorological, biological, geographical and military information. An explosion in this field accompanied man's progression into what is popularly called the space age.

The methods used depend upon the way in which matter absorbs, reflects, refracts or radiates electromagnetic radiation; each can be measured by airborne instruments or by scanners mounted on board space satellites. Conventional photography can detect only that part of the electromagnetic spectrum between long-wavelength ultraviolet

BELOW LEFT The Earth at half phase. A photograph taken from Apollo 12 and clearly showing the western edges of both North and South America and complex cloud patterns.

BELOW RIGHT The electro-magnetic spectrum, showing remote-sensing bands.

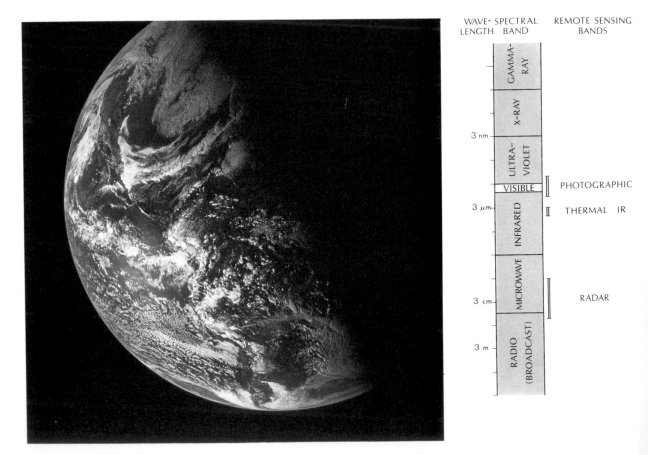

WAVE-LENGTH	SPECTRAL BAND	REMOTE SENSING BANDS
	GAMMA-RAY	
	X-RAY	
3 nm	ULTRA-VIOLET	
	VISIBLE	PHOTOGRAPHIC
3 μm	INFRARED	THERMAL IR
	MICROWAVE	
3 cm		RADAR
3 m	RADIO (BROADCAST)	

Skylab 4 photograph of part of Arizona. This shows Lake Mead (left-hand edge of frame) and the Colorado River. This runs first north and then east from the lake, incising itself into the countryside through the famous Grand Canyon, which exposes perhaps the most continuous section through the rocks of North America. Toward the top right-hand corner of the picture is the aptly named White Granite Gorge.

and short-wavelength infrared. Photographs are usually taken on panchromatic film, using special filters to minimize the effects of atmospheric scattering. Colour photographs may also be obtained and specially prepared colour infrared film has been developed for this.

One modern development is that of the multispectral scanner (MSS). This is a remotely operated system which employs a mirror to scan a swathe of ground beneath the spacecraft. The radiation collected is then split into discrete wavebands by a series of prisms, passed to one of a number of photocells and then recorded on multi-track magnetic tape. The stored information is periodically transmitted back to receiving stations on the Earth's surface.

A further method uses an energy source actually generated aboard an orbiting satellite. This has been successfully tried out with the Space Shuttle and involves the transmission of a radar pulse from the spacecraft. Information reflected from the target area is collected aboard the satellite and then recorded onto magnetic tape. This can subsequently be processed to give a visual image. Some very spectacular pictures have been obtained recently by this technique. One of its particular advantages over MSS data is that radar penetrates atmospheric clouds – which is why it has been extensively used to map the cloud-covered surface of Venus.

Much geological mapping is still based on study of photographs collected by survey aircraft. Usually this is arranged in such a way that each frame, photographed vertically down from the aircraft, overlaps that adjacent to it. In this way stereoscopic views of the terrain under survey can be obtained, and by using a mirror stereoscope to view overlapping stereo pairs of pictures, it is possible to view them in 3D.

With the advent of satellite photography and the introduction of

LEFT A high oblique view of the great Nile Delta and the Sinai Peninsula, taken from the spacecraft Gemini 4 on 4 June 1965. Note the northern forked end of the Red Sea toward the top right, formed in response to continental rifting in Tertiary times.

BELOW View of the volcanic calderas of the western Galapagos Islands, Ecuador. Taken by the imaging radar system on board the Space Shuttle Columbia, 1981.

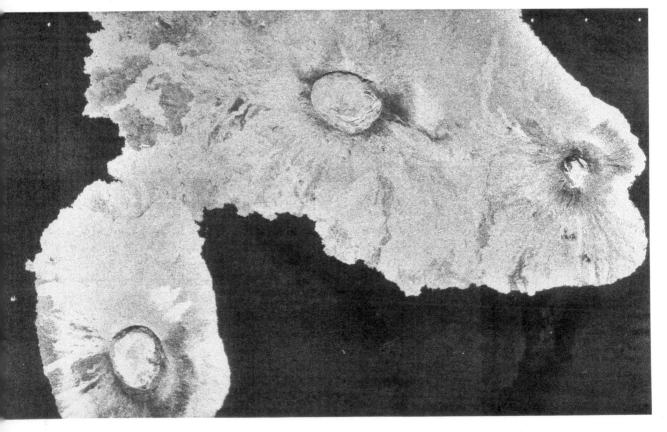

The Baluchistan Desert photographed by Landsat on 25 November 1972. This region lies where the borders of Pakistan, Iran and Afghanistan meet. The edge of a series of longitudinal sand dunes lies to the north (top right corner). The Tahib River runs across the view, while to the south are the folded and faulted Cretaceous and Tertiary rocks of the Mirjawa Range. Note the large dormant volcano Kuh-i-Taftan which lies close to the bottom left corner of the frame. This deeply dissected cone rises to a height of 4615 m.

optical-mechanical devices like MSS, much emphasis is being put on sophisticated methods of interpreting and manipulating the data by using computers. This is primarily because MSS and radar images are collected in digital form and can therefore be fed into a computer from magnetic tape, stored and studied in a variety of ways at the researcher's convenience. A large number of satellites orbiting the Earth now carry scientific payloads; of these, two are at present of particular interest to the geologist. These are called Landsat-5, which is the currently operating member of a series following four similar earlier spacecraft, and Seasat, which has recently provided fascinating information about the Earth's oceans.

Views of the Earth – 1

Landsat, Seasat and the Space Shuttle have all recently provided geologists with an immense amount of new data. However, several earlier missions should not be forgotten. These include the manned Gemini and Skylab flights and, of course, the Apollo missions which, besides taking photographs of the Moon and landing men upon its surface, also obtained a large number of images of the Earth.

Views of the Earth-2

A great deal of interest was aroused by the recent American Space Shuttle missions, not least among the reasons being that on board one very recent craft, Columbia, was an experiment known by the acronym SIR-A, or 'Shuttle Imaging Radar-A'. Although this part of the mission ended prematurely through instrument failure, sufficient imagery was obtained to prove the value of this technique.

Pulses of microwave radiation (1 mm – 1 m wavelength) transmitted towards the Earth can penetrate clouds and atmospheric haze, and can therefore reach the surface of the Earth. It will be re-emitted from the rocky materials it encounters in different ways, each particular material having its own radar 'signature'. The material properties that affect this signature are: (i) the surface roughness; and (ii) the di-electric constant – a measure of a material's insulating properties, which increases proportionally with the moisture content. Smooth surfaces appear dark on radar images, whereas rough ones (which randomly scatter the radiation) look bright. A good radar image will highlight the topography of an area and therefore enhance linear and other structural features of interest to the geologist.

Photograph of Kalpin Chol and the Chong Korum Mountains, Xinjiang, China. This radar image was obtained by the Space Shuttle Columbia in 1981. It shows striking linear features which are the result of folding and faulting of the rocks during late Tertiary and Quaternary times.

Views of the Earth-3

The electromagnetic radiation collected by photographic satellites originates in the Sun. The Earth does, however, emit radiation of its own, but this is of long wavelength and has to be collected by instruments specially built to measure energy in the far infrared. This thermal radiation can be either re-emitted solar energy; geothermal energy released in the vicinity of volcanoes; or man-made energy, such as that put out by power stations or in areas of urban population and industry. Thermal IR measurements are normally made at night so that the swamping effect of reflected solar radiation is minimized. Thermal images look rather different from normal ones and are extremely useful for vegetation surveys, for distinguishing differences in rock types and for analysing the thermal output of volcanic regions.

RIGHT Quantitative night-time infrared image of Hawaiian volcanoes, taken in February 1973. This is a specially processed image which shows the summit of Mauna Loa (left-hand side) and part of the Southwest Rift Zone with pit craters on its floor (running horizontally across the frame). The thermal characteristics of the different volcanic rocks are picked out by different colours, thus those with the lowest radiant temperatures are shown in black, the highest in white. Note the relatively high temperatures along the rims of the craters, due to these being built from relatively dense lavas which contrast with the more porous flows of the volcano's flanks and caldera floor.

BELOW RIGHT False-colour Landsat image of the region of Kilimanjaro volcano, on the border of Kenya and Tanzania. In this picture vegetation is picked out by reds and is seen to be concentrated on the flanks of the giant volcano of Kilimanjaro (5895 m), the neighbouring cone of Meru to the west and also Monduli mountain west again. Scarps entering the region in the northwest are part of the main rift faulting of the great Rift Valley.

Radar image of the Andaman Sea (Indian Ocean) acquired by the Space Shuttle Columbia in November 1981. This reveals in subtle detail large internal waves about 150 km southeast of North Andaman. They probably are generated by the strong interaction of tidal currents with topography on the sea floor. Once generated, the waves appear to propagate along the boundary layer between waters of differing density.

Seasat View of the Oceans

In 1978, the National Aeronautics and Space Administration (NASA) launched a satellite called Seasat. On board was a sensitive radar altimeter capable of making very accurate determinations of the height of the sea surface. After three months of operation this failed, but before doing so had surveyed the oceanic regions between latitudes 72° north and south. It provided geologists with what is called geo-tectonic imagery.

The measurements made, accurate to between 5 and 10 cm, can be processed to reveal striking images of the ocean floors. The basis for such a conversion is the knowledge that the sea height is directly related to the strength of the Earth's gravity field at sea level. Thus the water piles up in regions of high gravity and sinks down where the strength is low. The gravity field is at least partly dependent upon sea-floor topography, so the information obtained approximates to a topographic map of the ocean floor.

Seasat gravity map of the Earth's oceans.

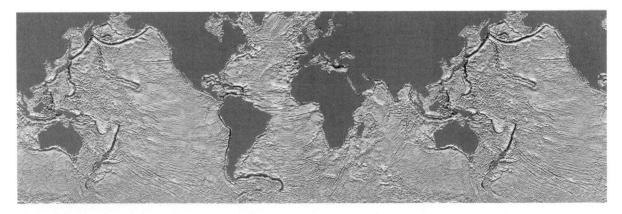

Part Three

PATTERNS OF EARTH HISTORY

The Earth has been in existence for at least 4700 million years, and during this time the heat engine has caused large-scale modifications of the crustal layer. New crustal rocks have been created; some have existed for a while, then been reabsorbed; others have found a place on the surface for much longer, although they may have been severely modified. In trying to understand the story of the Earth, geologists have to follow the threads of a giant web as they weave their way confusingly among the complexities of Earth history. One feature of this immensely long saga is its cyclic nature, similar sequences of events having occurred many times as Earth orbits the mother star.

Geologists have risen to the challenge in different ways, some using quite simple methods, others utilizing sophisticated technology; but whatever the method, the pattern of geological history must be sought in the rocks themselves. The rocks reveal a fascinating story, the bare bones of which we will now attempt to describe. This is not an easy task, and the reader must judge whether or not we succeed.

Eroded blocks of Silurian strata on a modern beach, West Wales.

10

RADIOMETRIC DATING

Elements and Isotopes

The chemical elements are composed of atoms which are held together by electrical forces. Each atom consists of a relatively heavy nucleus, consisting of positively charged protons and neutral particles called neutrons. Roughly 99.9 per cent of the atom's mass resides in this nucleus. Orbiting it are one or more negatively charged electrons – very light particles indeed – whose total negative charge is exactly balanced by the positive charge of the nucleus.

The sum of the protons and neutrons in an atomic nucleus is called the element's 'mass number', while the total tally of protons alone provides its 'atomic number'. Atoms having the same atomic number belong to the same chemical element. Those having the same atomic number but different tallies of neutrons are called isotopes. All elements have at least two of these, some have many more: for instance, mercury has seven natural isotopes.

Some elements have naturally occurring isotopes that are radio-active and these may decay spontaneously to lighter isotopes of other

Frontispiece to Murchison's *Siluria*, depicting the geological formations of part of northwest Scotland.

chemical elements. Because this disintegration takes place at a rate that is fixed for any particular isotope, it provides geophysicists with a very powerful method of dating certain minerals that commonly occur in the crustal rocks. This is called radiometric dating and it forms the basis of geochronometry.

Radiometric Dating

One of the geologically important elements that has a number of radioactive isotopes is uranium, another is thorium; there are also potassium, argon and rubidium. Using uranium as an example: the isotope uranium-238 decays by a somewhat complicated process to the non-radioactive isotope of lead, lead-206. In doing so the unstable uranium nuclei emit charged helium atoms, called alpha particles, electrons, called beta particles and short-wavelength X-rays, called gamma rays. It is these emissions that bring about the change. The rate at which decay takes place bears a constant relationship to the quantity of uranium-238 present and is expressed in terms of the isotope's 'half-life'. The half-life of uranium-238, for instance, is 4500 million years; another of its isotopes, uranium-235, which is also radioactive, has a shorter half-life of 713 million years. The term 'half-life' refers to the time it takes for one half the original amount of uranium-238 to decay to the 'daughter' element (in this case lead-206). The isotope uranium-238 makes up about 99.3 per cent of the natural uranium at the present day; the remaining 0.7 per cent being composed of uranium-235. The much shorter decay period of the latter suggests that the proportion of uranium-235 must have been far greater in the distant past than it is now. Uranium-235 also decays to a non-radio-active isotope of lead, in this case lead-207.

In a similar way, thorium-232 decays to yet another non-radioactive

Diagram to show a typical radioactive decay series, beginning with one isotope of uranium (U) and ending with another of lead (Pb).

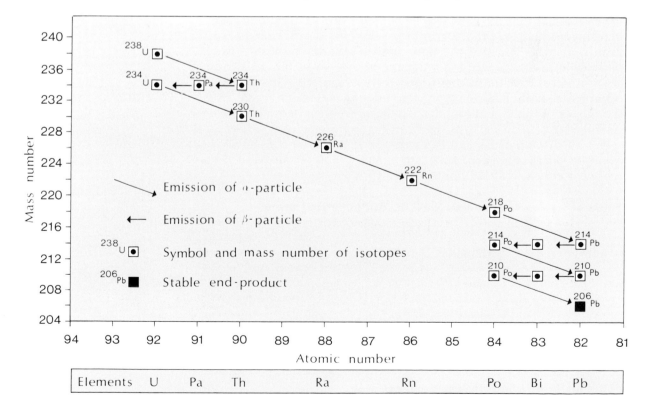

isotope of lead: lead-208. Thus we see that natural lead includes three radiogenic isotopes; there is also a fourth isotope of non-radiogenic origin, namely lead-204. A significant number of minerals contain uranium, or thorium, or both, and by using a complex instrument called a mass spectrometer, it is possible to find the ratio of radiogenic leads to the parental radioactive isotopes of uranium and thorium. The ratio provides the basis for calculating the age of the mineral.

Those minerals that do contain the above elements are all relatively rare, but another decay sequence occurs in the more common elements potassium, rubidium and argon, which are found in a wide variety of minerals. Thus age dating is carried out on parent-daughter pairs of elements: potassium-argon, rubidium-strontium and neo-dymium-samarium, as well as on pairs of argon isotopes and some isotopes of carbon. Together these various decay sequences provide the geophysicist with a powerful dating tool for rocks of different types and ages.

For success to be achieved with these modern techniques, the minerals themselves have to fulfil certain requirements. Firstly, they must be free of alterations produced by circulating solutions, since these might well have leached out a proportion of the measured elements, so skewing the ratios measured. Secondly, metamorphism can drive out some elements more rapidly than others and so particular care has to be taken where such activity has left its marks.

Where identical rocks have been dated by different decay sequences, there is usually a high degree of consistency in the derived ages. There are exceptions, of course, but over the years many of these have been explained. One example of an apparent inconsistency can be usefully cited here: a rock was dated by the potassium-40/argon-40 method. When used on a hornblende crystal in the rock, an age of 1000 million years was calculated. A mica crystal in the same sample yielded an age of 750 million years. How could this be? The solution lay in the fact that the rock was heated up during a period of metamorphism which occurred 250 million years after the rock crystallized from a magma. Because the mica has an atomic structure that is permeable to argon, while hornblende is not, all of the radiogenic argon was lost from the mica 750 million years ago, when it was heated up, at which point it again began accumulating radiogenic argon. Thus the timing of the metamorphic episode, as well as the rock's age, is established.

Not all discrepancies can be explained in this way but the above example shows how very useful radiometric dating can be, if used with due care and attention to detail.

Table of Isotopes commonly used for Radiometric Dating

Isotopes		Half-life of parent
parent	daughter	
uranium-238	lead-206	4500 million years
uranium-235	lead-207	710 million years
rubidium-87	strontium-87	4700 million years
samarium-147	neodymium-143	130,000 million years
potassium-40	argon-40	1300 million years
carbon-14	nitrogen-14	5730 years*

*This method is described in a later section.

Records of a Restless Earth

Records of the Earth's history are found among the crustal rocks, particularly those which are stratified. The elucidation of its past history by studying strata is called stratigraphy and many conventions attach to it. The basic unit for stratigraphic work is the 'formation': this is simply the name applied to a particular collection of strata which are sufficiently distinctive to form a unit for the purposes of mapping. Each formation will tend to contain its own particular fossil assemblage and, during the early days of stratigraphy, this was taken as sufficient evidence to assume formations extended laterally over vast distances, without perceptible change. The situation is not as simple as this, however, and it was realized in the late eighteenth century that the similarity of fossils found within the same kinds of sedimentary rocks might well be a reflection of environmental factors. In 1789, the famous French chemist, Antoine Lavoisier, published a number of diagrams which clearly showed he appreciated the effects that changes in the relative levels of land and sea had upon both sedimentary deposits and the fossils they contained.

Nearshore or 'littoral' deposits, formed in shallow turbulent water, tend to be coarse-grained and contain remains of organisms adapted to cope with these conditions. In contrast, offshore or 'pelagic' sediments, which form in deeper water, are relatively fine-grained and contain rather delicate bottom-dwelling organisms, as well as free-swimming or floating forms. It is easy to appreciate, therefore, that deposits of the same age may look quite different and also contain dissimilar fossils.

If we are to understand clearly what a series of strata represent in terms of past conditions at the Earth's surface, then we need to describe not just their vertical sequence but also their lateral distribution and variation. The lateral variations shown constitute what is termed a sedimentary facies which may be defined as the characteristics of a sedimentary rock that indicates its particular depositional environments. An example of facies variation is the passage from pebbles, through sand to silt and mud as you move progressively offshore from a modern shoreline.

A bore hole sunk vertically through a thick section of marine sediments will almost certainly reveal a varying sequence of rocks. Most of the variations encountered will reflect the effects upon sedimentation of minor fluctuations in the relative levels of land and sea. Broadly speaking, in marine sequences two kinds of pattern may be recognized: transgressive or 'onlap' sequences, which reflect a relative rise in the level of the sea compared to the land; and regressive or 'offlap' sequences, which are the result of a relative lowering of sea level.

Unconformities

Major punctuations in the stratal record are caused when a region of the crust previously receiving sediment is uplifted for a long period. This can happen when fold mountains are formed. The uplifted and deformed crust will be subject to erosion and gradually the older strata will be worn away and eventually may be resubmerged. Rock debris from the uplifted areas will be transported into the sea and there deposited as sedimentary rocks. At a later stage the ocean may encroach upon the eroded stumps of the older deposits and lay down sediments which show a marked angular discordance with the older rocks, creating what is called an unconformity. The famous Scottish

geologist, Sir James Hutton, predicted that unconformities should occur between older and younger rocks. In the two years following 1787, he found what he had been seeking. A major unconformity in Arran, Scotland, and subsequently the more famous one at Siccar Point, where older steeply dipping rocks were overlain with obvious angular discordance by flat-lying sediments of lesser age.

Angular unconformity between inclined Cambrian mudstones (below) and semi-horizontal Ordovician sandstones. The hammer is 30 cm long. Trwyn Llech-y-doll, Lleyn Peninsula, North Wales.

There are various degrees of unconformity, depending on what has come to pass during the period represented by the gap in the stratal record. Where there is no angular discordance between older and younger strata, though a significant time gap is known to exist, the term disconformity is used. Situations like that described by Hutton at Siccar Point, where there is a marked discordance between older and younger strata, are termed angular unconformities.

Unconformities, being boundaries between rocks, can be mapped across large areas. During such an operation care would be taken to establish the age of the rocks both immediately above and below the plane of unconformity, because only in this way is it possible to determine its age. Also, the time interval represented by the unconformity is necessarily greater at some places than at others; in some places it may even peter out altogether, to pass laterally into a conformable sequence.

The Cycle of Events

The end-product of the study of strata and also of igneous and metamorphic rocks is the elucidation of geological history over large regions of the Earth. Maps showing the lateral distribution of different rock types, the traces of unconformities and the position of dislocations such as faults, are the basis for description of both local and regional geology. Cross-sections through well-chosen points will aid an appreciation of the vertical as well as lateral variations the rock units show. When dealing with geology on the continental scale, it may be appropriate to draw up special maps showing facies variations, or even the thickness of particular geological formations that have great significance.

There are innumerable problems with reconstructing Earth's history, not the least being that of timing events. An unconformity may not be of the same age everywhere, nor may a thick sequence of sedimentary strata be, nor a series of lava flows. Geological processes usually act very slowly and they do not necessarily act universally. Thus although we know the Alpine mountain system was raised in Europe during Cenozoic times, we have no reason to think similar mountains were formed in, say, central Africa, at this time.

Despite all the obvious problems, it is possible to discern a cyclicity in the geological record. Similar sequences of events, taking perhaps hundreds of millions of years to complete, can be recognized. What is clear, however, is that we must pay attention not only to the sedimentary strata but also the magmatic and metamorphic rocks that occur alongside them; otherwise we may well miss many clues.

Orogenic Belts

Individual volcanic eruptions and earthquakes both disrupt rocks, but generally their effects are quite localized. They may, however, form part of a much longer phenomenon that produces the greatest upheavals, namely mountain-building or, as it is technically known, orogenesis. The Earth's fold-mountain chains are all expressions of a complex series of processes, including the burial, deformation and subsequent uplift of thick piles of sedimentary rocks, particularly those which formed near plate boundaries, and were caught up in the powerful forces unleashed there. Our planet accomplishes most of its crustal destruction and deformation at plate margins, as a result of which the active zones in which mountains are generated are most often linear in form. They are called 'orogenic belts'.

Associated with mountain-building episodes there is always volcanic activity; the crust is invaded by hot magmas originating either toward its base or in the mantle. During orogeny huge wedges of sediments which accumulated near continental margins may get carried down to great depths, there not only to be deformed, but also heated and strongly compressed, so as to become almost unrecognizable. The transformations produced come under the heading of 'metamorphism'. Magmatism and metamorphism both conspire to complicate the stratigraphic record, but they also provide vital clues toward understanding the workings of our planet. To try and interpret the geological record without reference to these processes would be foolish in the extreme, because we would be ignoring phenomena that are a direct result of the release of energy from within the Earth.

Over the years it has become clear that the development, evolution and eventual decay of successive mobile belts has shown a tendency toward cyclicity that is characteristic of Earth history. Periods of relative quiescence have alternated with orogenies that have dictated major changes in the shape, position and topographic features of the continents. Thus the development, evolution and eventual stabilization of successive groups of mobile belts can be used to characterize major slices of geological history.

11

INTEGRATING EARTH HISTORY

Early in the twentieth century, two European geologists, J. J. Sederholm – working in Finland, and Arthur Homes – studying the ancient rocks of Africa, came to the conclusion that the Earth's continental crust bore an ingrained record of successive orogenic belts, each of which had contributed towards its growth. The Phanerozoic history of western Europe shows that rocks belonging to three successive belts are preserved here; thus in northwest Europe, one developed during the Early Palaeozoic, while a second in southern and central Europe in Late Palaeozoic times, and a third evolved in the Mediterranean area during Mesozoic and Tertiary times. By studying the rocks associated with each belt, it is possible to recognize patterns that have been followed during other orogenic cycles which may have affected regions far removed from Europe.

Individual orogenies had their own very distinctive characteristics, these being directly related to the events which occurred during a specific period of geological time. However, there is a tendency for major orogenic events, on a worldwide scale, to be grouped together in time; furthermore within such groupings the events which take place, at least superficially, are of similar type. As an illustration of the latter, we can cite the Caledonian and Hercynian orogenic cycles, which were related to continental collisions and the accompanying closure of former oceans. In contrast, during the Mesozoic period, the principal cycles were a function of continental rifting and sea-floor spreading events.

Naturally, an orogenic cycle shows pulses of waxing and waning activity; so that the development, establishment and eventual degradation of an orogenic belt is an extremely slow and lengthy process. It seems that the life span of most major belts is of the order of 200 million years or more; it may be four times as long.

Grouping the Geological Cycles

The sequence of events which characterizes any of Earth's mobile belts has variously been termed a 'geological' or 'orogenic' cycle. Such cycles have recurred throughout at least the past 2500 million years of geological time and have overlapped in both space and time in a way that renders the record extremely difficult to interpret.

Orogenic cycles typically include extensive magmatic activity, both deep down and at the surface of the Earth. Such activity may peak at particular times, giving clusters of radiometric ages for the igneous rocks produced. After studying the distribution of igneous rock ages for all of the continents over most of geological time, some workers believe it is possible to see global peaks in igneous activity, notably between 2800 and 2600 million, 1900 and 1600 million, 1150 and

The early geological cycles.

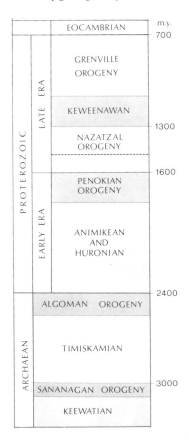

			m.y.
		EOCAMBRIAN	700
PROTEROZOIC	LATE ERA	GRENVILLE OROGENY	
		KEWEENAWAN	1300
		NAZATZAL OROGENY	
			1600
	EARLY ERA	PENOKIAN OROGENY	
		ANIMIKEAN AND HURONIAN	
			2400
ARCHAEAN		ALGOMAN OROGENY	
		TIMISKAMIAN	
			3000
		SANANAGAN OROGENY	
		KEEWATIAN	

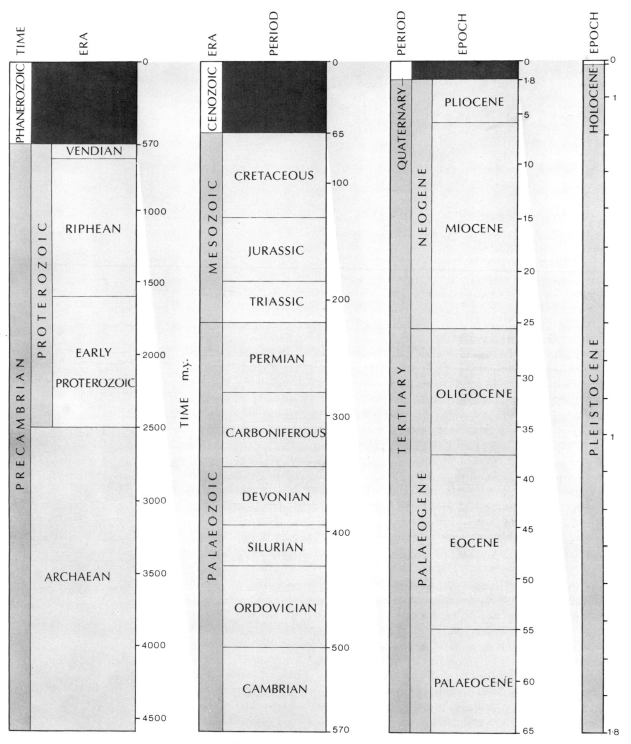

Geological time scale.

900 and 5000 million years ago. This is not to say that activity was absent between these periods, but they suggest that tectonic activity has not always been uniform. Geologists have recognized three gross subdivisions of time, (i) the Archaean, which includes rocks and events dating from before 2500 million years ago, (ii) the Proterozoic – cycles completed before about 1000 million years ago (although there is some continuing discussion about the precise dating here), and (iii) the Late Proterozoic to Phanerozoic.

Folded and cleaved rocks, Anglesey, North Wales. Buckle-type folds
affect the more resilient sandstones, while a pervasive fracture (cleavage)
passes through the less resilient shales (hammer).

Orogeny and Volcanism

We have seen that along certain active zones within the oceanic basins,
new basaltic crust is being produced. Evidence from magnetic striping
and radiometric dating indicates that the newly formed crust, together
with the lithosphere beneath it, will then slowly spread away from
both sides of the spreading axis at rates that vary between 1 and 10 cm a
year. In many large oceanic basins, although best seen in the Pacific,
the oceanic crust is being drawn down into the mantle, eventually to
be reabsorbed there, at depths of around 700 km.

The return of the oceanic crust to the mantle below takes place along
inclined zones, which reach the Earth's surface at deep trenches in the
ocean floors. These represent convergent plate boundaries where one
of the plates is being deflected downward beneath the other. The
inclined zone down which this underthrusting occurs is known as
a subduction zone. Friction between the adjacent lithospheric plates
within the zone gives rise to frequent earthquakes whose depth of
focus increases toward the overriding plate. The sloping, seismically
active plane is called a Benioff zone.

As the oceanic crust is carried back into the mantle, the sedimentary
rock covering its leading edge is mostly scraped off and highly deformed, while the oceanic material starts to be partly remelted at
depths of anything between 100 and 300 km. It may, however, continue to subside largely intact to greater depths, finally breaking up at
around 700 km. Because it is substantially less dense than the mantle
material surrounding it, the magma produced as the sinking slab melts
rises toward the ocean floor and is erupted as lava, which often
emerges from chains of volcanic islands, along 'island arcs'. Japan and
the Aleutian Islands are typical examples.

In some cases subduction of oceanic crust takes place directly adjacent to a continent, as is currently happening along the western side of
South America. In such a situation some of the magma generated as the
oceanic crust is thrust beneath the continental lithosphere rises
through the sialic layer and emerges from volcanoes built high up on
fold-mountain ranges such as the Andes, while some rises into the roots
of the mountains and cools slowly within large granitic intrusions.

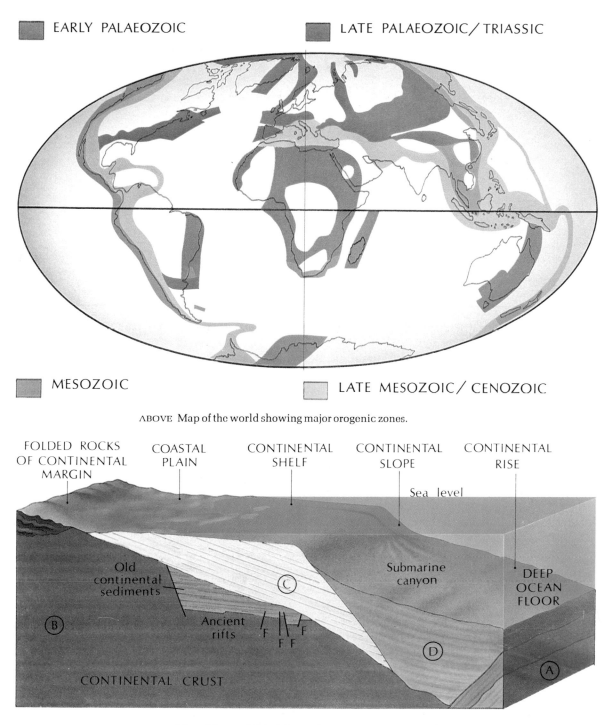

EARLY PALAEOZOIC

LATE PALAEOZOIC/ TRIASSIC

MESOZOIC

LATE MESOZOIC/ CENOZOIC

ABOVE Map of the world showing major orogenic zones.

FOLDED ROCKS OF CONTINENTAL MARGIN

COASTAL PLAIN

CONTINENTAL SHELF

CONTINENTAL SLOPE

CONTINENTAL RISE

Sea level

Old continental sediments

Ancient rifts

F F F F

Submarine canyon

DEEP OCEAN FLOOR

(B)

(C)

(D)

(A)

CONTINENTAL CRUST

ABOVE Map of the world showing major orogenic zones.

ABOVE Diagram to show development of large basins adjacent to a continental margin. A slab of oceanic lithosphere (A) is being subducted beneath a plate-bearing continental crust (B). Along the continental margin land-derived material accumulates in a sedimentary basin (C) beneath the coastal plain and continental shelf. In deeper water and further from the continental margin, marine sediments and volcanic rocks accumulate to produce a thick prism of sediments which underlie the continental rise and deep-ocean floor (D). Limestones, sandstones and shales typify the inner basin which develops in a stable environment. The less stable conditions experienced away from the continental margin immediately above the subducting slab are reflected in the accumulation of turbidites and submarine volcanic rocks typical of such environments.

It is perhaps now more apparent why orogeny and magmatism tend to go hand in glove. The series of concepts that comes under the heading 'plate tectonics' helps to explain how plate movements can account for the geological phenomena we see. Earth's most active zones occur at boundaries between lithospheric plates; this is where volcanism and seismicity are rife. Seismic activity without volcanicity occurs along transform faults (such as the San Andreas Fault) where plates are gliding past one another. Major mountain-building processes appear to be generated where plate convergence involves continental crust, as has been the case with the Himalayas, the Alps and the western Cordillera of the Americas.

Shields and Cratons

If we take a very broad look at the geological map of a continental mass, such as the North American continent, we at once see that it is built from different components. Extensive regions of the interior, especially in Canada, are quite flat and have remained more or less undisturbed since Precambrian times, at most having been warped or affected by up-and-down 'epeirogenic' movements. Such an ancient block of crust is called a shield. When analysed in detail, it is found that such shields consist of a number of Precambrian mobile belts that have been welded together and now form extremely resilient crustal units. Indeed ancient shields form the core regions of all of the present continents.

The highly metamorphosed character of typical shield rocks shows that at one time they must have been deeply buried, but hundreds of millions of years erosion, coupled with the epeirogenic movements, have seen to it that they have become exposed on the Earth's surface. The typical rocks of such regions are granitic and include highly metamorphosed types, called gneisses, together with an assortment of much-deformed sedimentary and volcanic rocks. Such an assemblage, often yielding very ancient radiometric dates, indicates there was extensive orogenesis during Precambrian times.

In other regions of North America the shield is covered by very old sedimentary strata which show only the most minor effects of either deformation or metamorphism. The sediments probably accumulated in subsiding basins formed in response to fracturing or downwarping of the ancient crust. These old strata, which range in age from Precambrian to Palaeozoic, provide us with vital records of the Earth's early geological history, including life. The sediment-covered and un-covered shield areas comprise what are often called cratons. Each of the Earth's continental landmasses has a cratonic core.

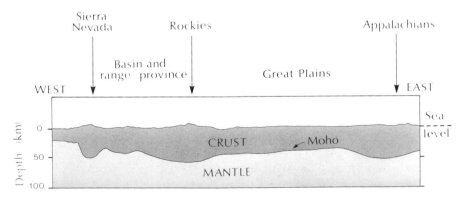

A diagrammatic section through the continent of North America, showing the general structure of a typical cratonic region.

Tightly folded gneisses, northwest Scotland. These ancient Lewisian rocks, originally sedimentary rocks, have been deeply buried during orogeny, metamorphosed and strongly deformed. They are typical of the core regions of an orogenic belt.

The Lifespan of an Orogenic Belt

Typically, orogenic belts are associated with continental cratonic areas. The evidence indicates that the deformed rocks arose from regions which, early in the evolution of the belt, received vast thicknesses of sediment. This apparently accumulated in thick wedge-shaped bodies which, in some cases, collected more than 10,000 m of debris. The sediment-laden troughs are often called geosynclines, a term that does not find favour with all geologists, but is frequently found in the literature. We prefer the term marginal basin. When two lithospheric plates converge bringing two blocks of continental crust together, the crust experiences severe compression and the accumulations of marine sediments are squeezed, as if by a vice, and deformed into great piles of folded strata which are often cut by low-angle faults, called thrusts. During such activity new mountains are thrown up along the site of the subducted ocean and during later stages, folded and faulted strata are thrust over onto the continental shields. Accompanying such movements are intrusions of granitic magma which solidify to form batholiths.

Once new mountains have been constructed, they are subjected to erosion, and during their formative years they will suffer severe attack, thick accumulations of coarse-grade debris being distributed both on the continental shield and alongside the new chain. Naturally, as the mass distribution changes with time, isostatic adjustments take place, and the mountains will continue rising for a long time – long after their formation. In some cases a chain may experience much later rejuvenation, the ancient deformed roots being eventually exposed high up amid the peaks of newer mountains. The Alps and the Appalachians provide us with two examples of this phenomenon.

Exactly how long does all this take? Naturally it is a very long time, but with the refinement in radiometric dating techniques seen over the last decade, and advances in our knowledge of the stratigraphy and deformation history of mobile belts, we can be more specific than that. The Alpine belt of southern Europe appears to have formed over a period of 200 million years and is still evolving; the Caledonian mobile belt, whose rocks are found in Britain, took much longer to evolve – about 450 million years; and the North American Cordilleran belt spans over 750 million years! Evidence suggests that some of the ancient Precambrian belts took even longer to evolve.

The Earth's Earliest Crust

What was the Earth's crust like? The Moon, having been virtually inert for the past 3500 million years, provides us with a far better record of its early crust than does the Earth. The most ancient Moon rock has an age of 4600 million years; so far the oldest terrestrial sample yields an age of only 4000 million years.

Meteoroid bombardment was severe during the first 800 million years of planetary history, a fact witnessed by the Moon's scarred crust, and there is every reason to suppose that Earth, too, once bore its marks. Being such a dynamic world, however, the events of this phase have long since been erased as the Earth's early crust was recycled.

As might be expected, the most likely places to look for samples of 'earliest crust' are the ancient continental shields, such as those that form the cores of the Americas, Africa, India and Australia. Until very recently the oldest dated sample came from a region of western Greenland called Isua, where gneisses exposed by the relatively recent retreat of the northern ice cap gave a radiometric date of 3824 million years. Associated with the Isua gneisses are a variety of highly altered sedimentary strata, chemically precipitated ironstones and cherts, together with volcanic rocks which clearly were laid down under water, indicating the presence of oceans at a very early date.

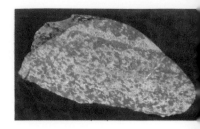

Earth's earliest rock? Photograph of a 4000 m.y.-old eclogite from the Roberts Victor Mine, South Africa.

Attention has also been focused on what are termed greenstone belts: greenish rocks first brought to light during exploration of the Witwatersrand gold belt of South Africa. The African greenstones are volcanic and altered ancient sedimentary rocks which have been engulfed in granitic rocks of the African craton. In the specific instance cited, an age of 3500 million years has been established for the lavas, turned green by metamorphic alteration of their original minerals. Within this belt of ancient rocks there are also volcanic and sedimentary rocks, produced during several cycles of volcanism and marine deposition. These rocks are overlain by a series of sandstones and conglomerates that appear to have been produced by the erosion of sialic crust; the greenstones, however, show no evidence of such an origin. The lack of such evidence makes them prime contenders for the label of 'primitive crust'.

The lavas, which have a very low silica content, are found interbedded with cherts; they often show 'pillow structure' – a common feature of submarine lavas. Such features show that oceans existed this long ago. A rather distinctive textural characteristic of the lavas, called spinifex texture, suggests the lavas were chilled very quickly and from temperatures approaching 1600°C: very much higher than those typical of modern basalts. This suggests significant differences between ancient and modern magmas, the precise details of which are still under discussion.

Very recently indeed, in August 1983, even older rocks were recovered from Precambrian volcanic pipes in southern Africa. These rocks, called eclogites, are rather unusual, dense, igneous-looking rocks which have been found in a variety of continental locations. The African samples, one from Tanzania and another from South Africa, yielded ages of 4000 million years. Geologists are debating whether these can be considered candidates for the title of primitive lithosphere, or whether the Witwatersrand lavas are stronger contenders. By the time this book is in print, perhaps we shall know the outcome!

Finally, recent dating of diamond-bearing rocks from South Africa establishes that the diamonds, while they were derived from the mantle, must have been incorporated in continental crust by 3500 million years ago. From what we know of the depths at which diamonds form, this suggests that even as long ago as that, at least 120 km of lithosphere had evolved and become distinct from the mantle beneath.

12

THE EARLY CONTINENTS

Pangaea – Supercontinent

Precambrian history is less easy to interpret than younger periods; for a start, the rocks are often much deformed and then there are no fossils to help us. For some time after it was formed, Earth was lifeless. Even when life began in the warm oceans, it left no detectable remains. Nevertheless, it seems that the ancient rocks of the continental shields date back at least 3500, probably 3800 million years; so there were continents at this early stage. From our studies of these old rocks, we believe the early sialic layer had a thickness of between 25 and 40 km compared with the present thickness of between 10 and 70 km. Furthermore, the continents may have been quite large: the total area of Precambrian North America appears to have been at least half as large as today's continent. The same may hold true for the others.

When we come to consider the origin and evolution of continental crust, we find there are conflicting opinions. It is possible that most was created very early on and was simply reworked by later orogenies. Yet it seems rather more likely that the continents grew slowly, with new crust being added to the margins of the existing shields. Whatever the method, there seems little doubt that the crust itself was produced by the chemical differentiation of the upper mantle.

Rocks from all of the shields have been dated by the radiometric

One possible reconstruction of the supercontinent of Pangaea. The ancient continental cores or 'shields' are shown in yellow.

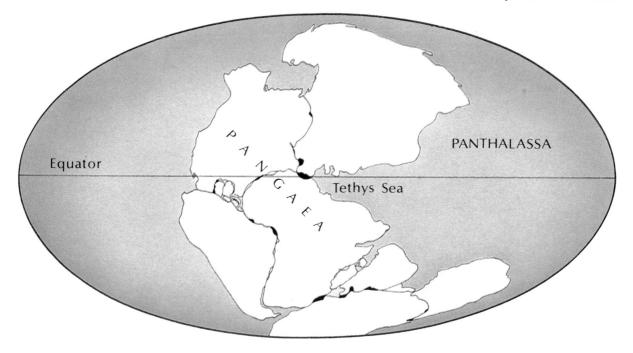

method; the ages range between 2500 and 4000 million years. Generally speaking the most ancient rocks are rather restricted, and it may be that the first major continent-forming episode started about 2800 million years ago where we find a cluster of radiometric dates.

We know a great deal more about the more recent rocks than we do about the Archaean and Proterozoic rocks, yet with the advent of palaeomagnetic methods, we started to understand how the continents had moved about during the long march of time. Were there several large continents during Proterozoic times? We cannot be sure; there may at times have been only one, a supercontinent whose late Proterozoic successor has been named Pangaea. Another alternative is that there were two principal landmasses: Gondwanaland, built from those shields found at the cores of the present southern continents, and Laurasia, made up of the northern shields. Stratigraphic methods alone cannot tell us the complete story; but when we draw on palaeomagnetic data, isotopic dating and structural studies, we can begin to form a fairly confident idea of what happened during these early stages of Earth's evolution.

Palaeomagnetic information suggests that there were, in the Late Proterozoic (1000 to 600 million years ago), five independent continents, corresponding to what we now call North America, Europe, Siberia, China and a continent called Gondwanaland (which no longer exists as an entity). Proterozoic 'North America' included much of the present continent plus northwest Britain, west Norway, Spitsbergen and possibly parts of northeast Siberia. Earlier (1100 million years ago), North America and Gondwanaland may well have been part of a single continent; and even further back, say 1500 million years ago, these two and Eurasia may have been combined, bringing us back to the notion of a supercontinent.

If this Proterozoic Pangaea did exist, and most of the evidence points that way, it did not persist. There were periods when Gondwanaland and North America split apart, only to re-unite. Mountain building, rifting and the opening and closing of oceans have all played a part in our world's past history; this is a pattern we shall now unfold.

Ancestral Continents

We do not know the precise shape of the early continents; but by using all the methods at our disposal, we can at least draw up a fairly reliable idea of the general picture. We shall start with the shield areas of North America and Eurasia, often grouped together under the heading Laurasia; then we shall look at Africa, South America, Antarctica, India and Australia, which are the remnants of the ancient Gondwanaland.

Ancient North America

Ancient North America was the direct predecessor of the modern continent. It has several components: a triangular-shaped block of crust called the North American craton, and two much younger belts of fold mountains – the Appalachians and the western Cordilleras, together with fold belts in east Greenland and northern Canada. The northern part of the craton comprises the Canadian Shield, where we find Archaean rocks, and similar rocks extend over all of the craton to the south, but are there blanketed by younger strata. A feature of the cratonic core is that it has been a stable region for at least 2000 million years.

The Canadian Shield has been carefully studied, even though parts

The main orogenic events to affect the North American continent during Precambrian times.

EAST GREENLAND MOBILE BELT
0.25 - 0.35 b.y.   0.3-0.4 b.y.

\> 1.8 b.y.

3.6 b.y.

1.2-1.5 b.y.

1.6 b.y.?

2.4 - 2.6 b.y.

CANADIAN SHIELD

CORDILLERAN MOBILE BELT

0.8-1.2 b.y.

0.3 b.y. - present

1.6-2.0 b.y.

2.3 - 2.7 b.y.

1.6 - 1.9 b.y.

1.2 - 1.5 b.y.

0.8-1.1 b.y.

APPALACHIAN MOBILE BELT
0.2 - 0.4 b.y.

LIMIT OF CRATON

Map of the ancient craton of North America showing the various 'age provinces' as established after radiometric dating of rocks collected from different regions. In the north the shield is exposed but further south it is covered by younger strata. Mobile zones border the edges of the craton.

of it are decidedly awkward to reach. Pioneer work was carried out during the mid-nineteenth century in the regions of Lake Superior, Minnesota and Ontario. The main problem with this terrain is that it is structurally very complex. Most of the rocks are highly deformed, and also metamorphosed; this makes them difficult to interpret. The oldest part of the shield lies in western Ontario but extends into Montana and Wyoming in what is now the United States; but in Archaean times, Greenland was joined to this landmass, and it is there that we find the oldest rocks of all, those of Isua. The ancient continent also included fragments of northwestern Britain.

Within the shield region there are two main assemblages of rocks: granitic rocks, including gneisses, and then volcanic and sedimentary rocks that formed in greenstone belts. During Precambrian times the shield was deformed by at least four episodes of crustal upheaval: one occurred about 3000 million years ago, another around 2400 million years ago, a third at 1600 million years, and a fourth between 1000 and 700 million years ago. Each episode gave rise to orogenic belts; this led

85

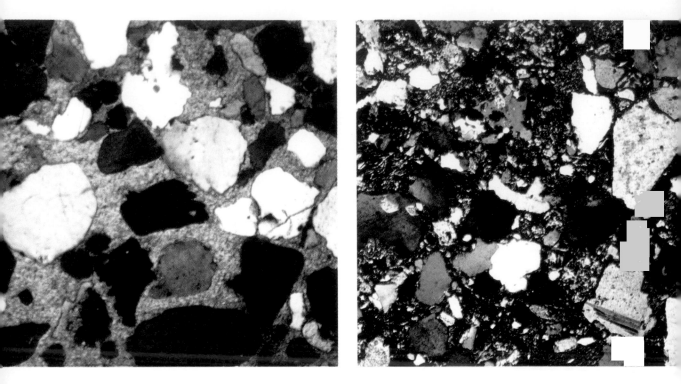

also to rifting, and in some cases to collision of lithospheric plates.

After each upheaval, new crust generated within the mobile belts stabilized and became welded to the perimeter of the existing continent. The large-scale rise of granitic magmas into the unstable regions helped to consolidate this process. Granitic rocks, being of low density, tend to 'float' on the denser lithosphere beneath; the greater the amount of low-density granitic material added to a continent, the more difficult it becomes for it to be subducted under the oceanic crust; thus it remains relatively undisturbed for long periods of time.

The sedimentary strata found in the Archaean greenstone belts are termed 'immature' and are predominantly what are called greywackes. By this term we mean that they contain a large proportion of volcanic fragments and were not simply produced by the destruction of pre-existing sial. They seem to have been laid down in the primaeval oceans during the unstable periods, and provide evidence for abundant volcanic activity. Not all of the Precambrian sediments originated in this way, however, for the ancient strata covering the craton away from the Canadian Shield evidently were derived from sialic crust. These rocks, said to be 'mature' sediments, are composed of minerals like quartz and feldspar which are more likely to have originated from granitic material; the way in which the grains occur in the rocks strongly suggests they were deposited in shallow, agitated water. There is thus every reason to suspect that these rocks were deposited under extensive shallow seas that invaded large regions of the ancient continent.

Very late in Precambrian time, thick layers of marine and sedimentary rocks accumulated along the continental perimeter. These rocks are now found in the Appalachians, the Western Cordillera of North America, east Greenland, Norway and northwest Scotland. One suggestion is that these rocks collected in down-faulted trenches (called graben) which were generated when ancient North America broke away from ancestral Europe.

ABOVE LEFT A mature sandstone as seen under the microscope in cross-polarized light. Grains of clear quartz are embedded in a cement of crystalline calcite.

ABOVE RIGHT An immature turbidite sandstone composed of a mixture of quartz and feldspar grains, volcanic and other rock fragments set in a matrix of fine-grained mud fragments.

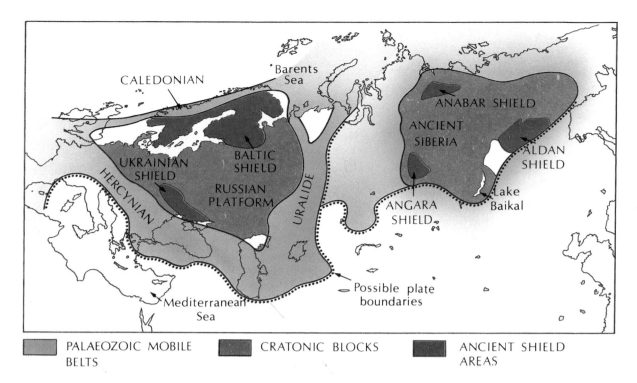

PALAEOZOIC MOBILE BELTS

CRATONIC BLOCKS

ANCIENT SHIELD AREAS

Map to show position of main shield areas in ancient Eurasia. The Russian Platform and Ancient Siberia represent cratonic areas where the shields are covered by later sediments. The positions of Palaeozoic mobile belts are shown in relation to the Precambrian cratons.

The Cratonic Rocks of Eurasia

Precambrian rocks are exposed in Scandinavia, the Baltic regions, the Ukraine, Britain and in parts of southern Europe; there are also outcrops in Arctic Russia and Siberia. Remnants of ancient crust are found, too, in eastern China, Manchuria and Korea. Various bits of evidence suggest that the Precambrian rocks extending from eastern Newfoundland to Massachusetts were a part of ancient Europe in Archaean times. The story of Precambrian times in this continent largely concerns the splitting apart and subsequent reunion of ancient Europe, Siberia and China, while the European continent also came into collision with ancient North America at times.

Radiometric dating tells us that the gneiss outcrops of the Ukraine and northeast Baltic Shield are between 3000 and 3100 million years old. Overlying these rocks, with strong unconformity, are layers of intensely deformed and metamorphosed volcanic and sedimentary rocks; these in turn show evidence of having been invaded by granite magmas about 1600 million years ago. The Archaean rocks appear to be remains of the old European and Siberian landmasses.

The Baltic Shield

The Baltic Shield is undoubtedly one of the most closely studied regions of the crust. It can be divided into three parts: a large region of Scandinavia and the Ukraine occupied by mobile belts of Early Proterozoic age; regions of older gneisses which lie in the north and east; and a western belt where an ancient mobile zone has been incorporated into a much younger orogenic belt known as the Caledonides. The first of these regions has extensive sedimentary and igneous rocks that are between 1700 and 2000 million years old. They are thus younger than the gneisses that outcrop further north; these have yielded an age nearer to 2500 million years.

The sedimentary strata exposed in Scandinavia and the eastern

Ukraine are over 500 m thick, and include quartzite, marble, schists and distinctive iron-rich rocks that are often found in ancient terrains. West of them we find metamorphosed shales and volcanic rocks that are presumed to have accumulated in a rather unstable marine environment. Around 1700 million years ago the more westerly parts of the shields were involved in an episode of folding and metamorphism, and were intruded by granites. One suggestion is that this upheaval was due to a collision between ancient North America and Europe.

Between about 1700 and 1000 million years ago, what is now Finland and Sweden was invaded by swarms of basaltic dikes. Dike formation is normally associated with the stretching of the crust, the rifts formed by the tensional stresses providing ideal channels for the movement of fluid magma. The stretching may have been the start of separation between the old North American and European continents; but later in the Proterozoic, a chain of new mountains arose at the western edge of the Baltic Shield. Such structures are not typical of tension – indeed they form under compression; clearly there was a change in the situation, at least in the west. The most likely explanation for the compressive episode is that there was a collision between ancient Europe and either Gondwanaland or ancestral North America.

Late in Proterozoic times, thick volcanic and sedimentary deposits accumulated in marine troughs which developed on the continental perimeters; these were deformed during Palaeozoic times. During Late Proterozoic times, the ocean that separated Laurasia from Gondwanaland extended from Nova Scotia through Newfoundland and central Britain, to Norway, Spain and southern Germany.

The Baltic Shield showing the various structural components. The Svecofennide mobile belts were stable by Early Proterozoic times. They sit west of older Precambrian terrain and east of younger belts, including the Caledonide mobile belt, which is of Palaeozoic age.

Ancient Siberia and China

There are not very many exposures of ancient rocks in Siberia, but sufficient gneisses are seen for us to realize that there are a number of shields and that these formed the core region of an early continent. The most northerly of the Siberian shields dates back 3600 million years. These may have formed part of a vast cratonic block that existed in very early times. In more recent times, rocks that had accumulated in orogenic belts were crushed against its margins, as continental collisions occurred.

Small shields are also found in Manchuria, east China and Korea; although our knowledge of these is not great, it seems, therefore, that there may have been a number of ancient Asian continents early on. Radiometric dating of these rocks is not far advanced, but those dates that have been obtained suggest ages of 2500 million years for the oldest shield.

Large tracts of metamorphosed Precambrian rocks occur along the southeastern and southwestern margins of the Siberian craton; they were formed in mobile zones which developed between 1600 and 800 million years ago. These Proterozoic rocks provide us with clear evidence that events in Eurasia were not dissimilar from those affecting North America at this time.

Gondwanaland

All methods of investigation agree in showing that the pattern of events in the ancient landmass called Gondwanaland was rather similar to that in North America and Eurasia. Periods of stability alternated with phases of mobility during which narrow zones of crust suffered substantial upheavals. Not only did Gondwanaland show movement with respect to Laurasia, but it seems that the ancient shields from which it was composed periodically broke apart and then came together again. Periods of unrest in Archaean times occurred 3600, 3000 and 2700 million years ago. The oldest rocks found among the crust of Gondwanaland are eclogites and have recently been collected in Tanzania and South Africa. Isotopic dating gives them ages of about 4000 million years. They may turn out to be the oldest primitive materials on Earth.

It is also in Africa that we find very large regions of ancient stable crust, although this does not mean to say that stability prevailed everywhere. Indeed in parts of Africa, as well as in what are now the continents of India and Australia there is abundant evidence for tectonic activity between 2700 and 2500 million years ago. The remains of ancient fold mountains produced at this time are now exposed on the surfaces of the modern continents.

At the beginning of the Proterozoic, quite thick volcanic and sedimentary deposits were laid down in southern Africa; this was also a time of intense rifting. Basaltic magmas rose into thousands of dikes which transect the crust and which also formed in Ghana, where they trend in a west-east direction and have an age of 2400 million years. Maybe this activity was connected with the splitting apart of Gondwanaland and Laurasia at this time. Certainly there is evidence in support of this and the most likely place for the original collision zone would be along the line of the modern Anti-Atlas Mountains of northern Africa.

In the Transvaal in South Africa, the geological sequence includes strata thought to be laid down under glacial conditions, about 2300

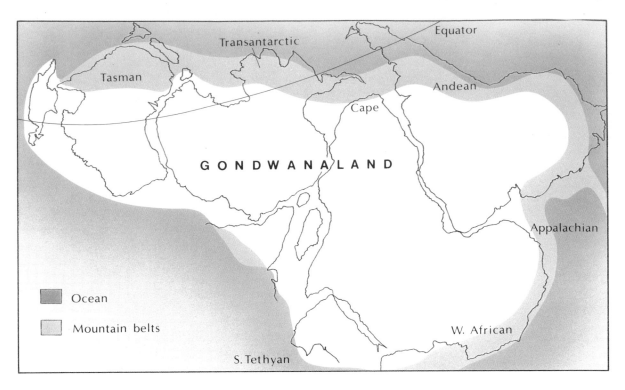

million years ago. Palaeomagnetic data show that when they were formed, this part of the continent lay within 30 degrees of the South Pole, and it follows that much of Gondwanaland must have been undergoing glaciation. Above the glacial formations are volcanic rocks; these are about 2000 million years old. Interestingly, similar sequences were being formed in each of South America, India and Australia at this time. There were great upheavals in both South America and Africa at about the same time as the volcanics were accumulating. It is possible that one of the early continents was plunging beneath the other along a subduction zone; or perhaps there were collisions between smaller 'microcontinents' that formed part of the major cratons. Whatever the details of these events, it does seem that around 1800 million years ago North America and Gondwanaland collided, producing a new fold-mountain chain.

Later in the Proterozoic they split apart again, and there was intense rifting about 1300 million years ago; similar events affected ancestral Europe, Siberia and North America. Later still, thick sequences of boulder deposits and limestones, thought to be glacial in origin, accumulated in regions as far apart as South Africa, Antarctica, Australia and northeastern South America. Palaeomagnetic evidence indicates that at this time, central Africa and northeastern South America sat on the Equator, while almost the whole of the African shield lay within 20 degrees of it. Although Africa was close to the Equator 600 million years ago, parts of it remained glaciated. Ice sheets clearly were not confined to the poles at this time in Earth's history.

More about Ancient Africa

Africa has had a complex history; basically it consists of a huge platform of ancient crystalline rocks which are partially covered by younger rocks. Much of this platform is built from Precambrian gneisses but there are Palaeozoic rocks as well. Much younger fold mountains occur in the north on the site of the Atlas mountains, and there is a Palaeozoic fold belt at the other extremity of the continent, in

Reconstruction of the continent of Gondwanaland, showing the ancient shield areas which now form the continental cores of Africa, South America, Australia, Peninsular India and Antarctica.

90

Ancient Africa, showing Precambrian shield areas. Each of the principal shields has a structural grain that strikes west to east, and is composed largely of granitic igneous and metamorphic rocks.

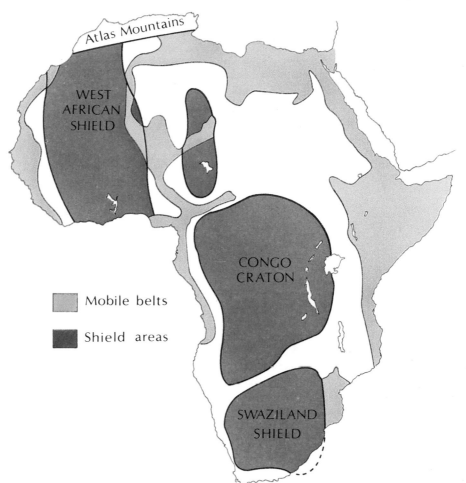

Cape Province. The ancient craton remained stable even while the Palaeozoic mountains were being produced. The famous East African Rift Valley which slices its way through the eastern side of the continent is a much younger structure, although earlier fractures had opened here as early as Precambrian times.

The principal regions of Archaean rocks occur in the Transvaal, Zimbabwe and Tanzania, Angola and West Africa; these have a west-east structural 'grain' and occasionally are enriched in valuable minerals such as gold. In many ways they are comparable to the ancient rocks of the North American Superior Province, but the Transvaal rocks in particular are of great interest, for they include extremely old sedimentary and volcanic rocks that have remained essentially unaltered since 3400 million years ago.

In each ancient block there are complicated interrelationships between sedimentary rocks, volcanics and granitic rocks. As elsewhere, gneiss terrains and greenstone belts occur here – and there is a general feeling among geologists that the same geological processes operated in Africa as in Canada and Eurasia. In broad terms it seems that during Archaean times there was a 'basement' of gneissose rocks upon which and against which younger volcanics and sediments accumulated, particularly during orogenic episodes; the latter arose where there were weaknesses in the existing granitic crust. One of the most prominent of these unstable areas separated the ancient gneisses of the Transvaal from those of Zimbabwe; this is called the Limpopo Belt. It appears to have been the site of long-lived crustal activity at least until 2000 million years ago.

About 1700 million years ago there is evidence to suggest that there was an episode of fragmentation in Gondwanaland. The North African Anti-Atlas Mountains contain sediments and volcanics that clearly are of deep-water origin and presumably were produced when this part of the supercontinent began to fragment. Twelve hundred million years ago, the various components of Gondwanaland appear to have re-united, but Late Proterozoic deep-water sedimentary rocks exposed in the Atlas Mountains show that 1150 million years ago Gondwanaland and Laurasia were once again rifting apart.

Peninsular India

India is best considered as two regions: the triangular cratonic block of peninsular India and Sri Lanka (Ceylon), and the Himalayas. The former has remained fairly stable for at least 500 million years; the latter region has been subject to upheavals throughout Phanerozoic time. The Indian craton collided with the Asian landmass in relatively recent times. The ancient craton has rocks which range in age from 2700 to around 500 million years. In the north, old cratonic rocks disappear beneath the vast sheets of debris which have been shed by the Himalayan chain. Much of India is composed of granitic gneisses at least 2000 million years old, but these rocks tend to be strongly metamorphosed and are often difficult to interpret. The largest block of ancient crust is situated in southwest India, and there are other size-able outcrops between Bombay and Delhi.

Several narrow orogenic belts cross the craton, these ranging in age between 2000 and 1200 million years. They run along the east coast southwest from Delhi and one belt extends inland from Calcutta; this is of extreme economic importance as it contains very substantial

Peninsular India showing the blocks of ancient crystalline rocks which lie to the south of the Himalayan mobile belt. A large region in the west of the peninsula is covered by younger basaltic lavas of the Deccan Traps.

deposits of iron ore. Among the rocks exposed are the very interesting 'charnockites', which show evidence of having been formed deep down in the lithosphere.

In the northern part of the peninsula is a vast region of Precambrian volcanic and sedimentary strata whose precise age remains somewhat in doubt. Structures produced in these rocks by ancient currents show that the quartzose sandstones and shales were evidently laid down in shallow water, along with some limestone seams. The current structures also suggest they were derived from land to the southeast, and it may be that they originated on a stable platform which existed soon after the old craton became stabilized at the end of the Proterozoic.

Antarctica

Although the Antarctic is 99 per cent covered by ice, geologists have been learning much about its geology in recent years. This has, of course, been the result of geophysical work. In the west of the continent there is a relatively young mobile belt, but to the east the ice is underlain by a cratonic block that is about 35 km thick. This east Antarctic craton includes both gneisses and charnockites, many of which have been dated radiometrically. The oldest dates so far obtained cluster around the 2 billion year mark.

In coastal east Antarctica many rocks give ages that centre on the period 650 to 400 million years ago. Here we appear to be looking at older rocks that have been reworked more recently. Near the western limit of the craton, in the Transantarctic Mountains, rocks of similar age have been found. Some of these rocks apparently have been derived from sediments laid down in a marine trough of Late Precambrian-Early Palaeozoic age, thus the history of this area goes back considerably further than the radiometric dates suggest.

South America

A brief survey of a topographic map of this continent reveals that it can be divided into three principal regions: the western cordilleras – the Andes; a broad region of plains country that lies east of the Andes and narrows eastwards into the Amazon basin; and the eastern high plateaux, separated into Guyanan and Brazilian components by the River Amazon. Not surprisingly there is a close connection between this tripartite division and the underlying geology.

The oldest rocks are found on the Guyanan and Brazilian Shields, but there are also smaller ancient blocks to the south in Argentina. In each of the main shields there are Archaean rocks which include gneisses, and there are also considerable thicknesses of both volcanic and sedimentary rocks. In Guyana the old gneiss basement is covered by younger pink and red sandstones, shales and volcanics in which contentious 'life forms' have been located. There are also many intrusive sills and dikes. These have been dated, fixing a minimum age for the Archaean rocks of between 1800 and 2000 million years. Other dated horizons yield greater ages (2500 million years plus). Nearer to the Atlantic coast the rocks have been deformed by a later Palaeozoic orogeny, during which substantial bodies of granitic rocks were emplaced. After this upheaval, the old craton was apparently domed up, to form a broad region of basins into which younger sediments were poured. These accumulations have remained undisturbed ever since.

The Brazilian region is underlain by a cratonic block, the ancient

South America. The main shield areas lie to the east of the great Andean ranges. The largest of the ancient blocks are the Guyanan and Brazilian shields whose rocks date back to at least 2500 m.y. ago. The huge Amazon basin separates the two main shield regions.

gneissose basement extending as far south as the River Plate. Both Archaean and Proterozoic rocks are exposed, the oldest being found in the north and east (2300 to 2500 million years). Along the coastal belt is a zone of deformed sedimentary and volcanic rocks, but these become less disturbed inland. On top of the ancient craton are several regions of Late Precambrian clastic sedimentary rocks, some of which are glacial in origin.

Australasia

The Precambrian rocks of Australia are widespread and occur in a large cratonic region which underlies the central and western part of the continent. Younger rocks cover much of the craton, but exposures are good in many areas. This is particularly true of the Yilgarn and Pilbara 'blocks' of Western Australia, which are separated from one another by the Hamersley Belt, a west-east zone of Precambrian orogenic activity. The rocks covering the ancient craton are of particular interest because they provide evidence for almost continuous sedimentation over a 1500 million year period. As with other cratonic regions, granite-gneiss-greenstone assemblages are typical. Some of the volcanics and sedimentary sequences are amazingly almost entirely undeformed, even though they are of such immense age. These appear to have accumulated in cratonic basins that remained largely unaffected while orogenic movements deformed rocks of similar age in adjacent regions.

In Western Australia, rocks of volcanic greenstone belts have yielded radiometric ages of around 2700 million years in the Yilgarn Block and 3000 million years in the more northerly Pilbara Block. Numerous

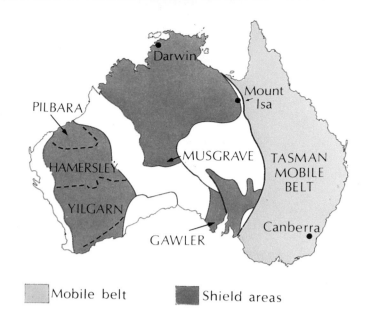

RIGHT Australia. Vast areas are built from Precambrian rocks, many of which are of Archaean age. Some of the rocks are of great interest since they have remained undisturbed since they were deposited, at least 2500 m.y. ago, and provide geologists with a record of cratonic deposition that is 1500 m.y. long.

BELOW Precambrian banded iron deposits of the Hamersley Iron Formation, northwest Australia. These rocks, made of chert, are over 2000 m.y. old, and were deposited along the margins of the ancient craton in sea water. The occurrence of these peculiar banded iron formations on almost every ancient continent is in itself interesting but, coupled to this, is the fact that all are between 2300 and 2000 m.y. in age. It has been generally agreed that they are chemical sediments which became oxidized after deposition. The oxidation process appears related to the photosynthetic activities of ancient organisms.

volcanic-sedimentary cycles are represented in these thick sequences, showing that the early history of the continent was not a quiet one.

Igneous activity also affected northern Australia during Early Proterozoic times. Radiometric ages of between 2000 and 1700 million years are typical for plutonic rocks from a wide range of localities. A number of younger igneous episodes occurred in mobile zones developed between the more stable blocks between 1700 and 1000 million years ago, suggesting that the continent may have been disrupted by repeated movements of smaller 'microplates' within the main continental mass.

The Hamersley Belt, with its west-east trend, saw the accumulation of one of the most remarkable successions to be found on continental crust. Resting on the Archaean basement is a 13 km thick sequence of volcanic and sedimentary rocks which evidently accumulated in a rapidly subsiding marine trough. Among the succession is a series of banded iron-rich rocks which, unlike the rocks above and below them, quietly precipitated on an undisturbed sea floor. The iron formation extends over at least 140,000 km^2, thinning away from the central regions, suggesting that it was laid down in an enclosed basin that remained undisturbed for a considerable time.

One of the most economically important regions of Precambrian rocks is that centred on Mount Isa, Queensland, where during the Late Proterozoic there was a period of extensive mineralization. This produced the famous deposits of copper, silver, lead, zinc and uranium for which this region is justly famous. Late in the Proterozoic, glaciation affected the area, as it did elsewhere in the world, and in several localities boulder-bearing deposits are seen to rest on a striated, glacier-worn surface. Several episodes of glaciation appear to have taken place; radiometric dating of the sedimentary rocks gives 875, 740 and 610 million years as the dates for these events.

The First Signs of Life

In its earliest days the Earth was quite unsuited for life, at least of the type we understand. As we have seen, the original atmosphere was stripped away, and a secondary one was formed from gases expelled from the interior. This atmosphere was rich in carbon dioxide, together with water vapour, hydrogen sulphide, carbon monoxide and hydrogen. There was very little free oxygen, so that a modern animal,

Stromatolites in position of growth, Shark Bay, Western Australia. They are believed to be the remains of blue-green algae and are among the oldest biogenic deposits, many being over 3500 m.y. old.

even a very lowly one, would have found conditions intolerable. However, there was one bonus. Although the Sun had only three quarters of its present luminosity, the concentration of carbon dioxide in the atmosphere acted rather like a greenhouse, keeping in solar radiation so that the surface temperature remained high.

It is not easy to say just when life on Earth began. Today, of course, it is everywhere, and is found in the most unlikely places; over one and a half million species of animals and half a million species of plants are recognized, and new discoveries are being made at the rate of thousands a year. But on the primitive Earth the environment was overwhelmingly hostile, and this must have been the case for almost a thousand million years after the Earth formed.

It is possible that the first signs of life can be tracked down in rocks in the Isua region of Greenland. These date back to 3800 million years and contain carbon. Of the two more abundant isotopes, carbon-13 and carbon-12, there seems to be rather less carbon-13 than in inorganic materials and this is a characteristic of biological activity.

Whether or not this indicates the presence of living organisms or not is a matter for dispute. However, we can be more sure about stromatolites, structures made up usually of calcium carbonate precipitated or accumulated by blue-green algae. Algae, of course, are unquestionably 'plants'; seaweeds are modern examples. Stromatolites date back 3500 million years; the oldest examples come from western Australia – from a place so remote that it is called 'North Pole'!

More positive evidence comes from the same region, but from somewhat younger rocks, formed 2800 million years ago. In these are very fine filamentous structures which are usually called pro-algae, and which almost certainly are biological. In the Lake Superior region of Canada we find slightly more complex plant-like forms, some of them not unlike the blue-green algae of today. Conditions on the Earth had evidently altered by that time. There was still very little free oxygen in the atmosphere, but at least there was enough to produce a layer of high-altitude ozone, which shielded Earth's surface from harmful shortwave radiations from space. Without this, life on Earth may never have evolved.

Then, in rocks of about 1400 million years old, comes the earliest occurrence of cells of more advanced type, called eukaryotic. Relatively quickly on the geological scale, much larger cells appeared. By 670 million years ago the oxygen level had reached perhaps 7 per cent of today's value. This was the start of what is called the Phanerozoic period when abundant, albeit primitive life had developed in such a way as to enable it to withstand the rigours of geological processes and become fossilized.

13

THE DRIFTING CONTINENTS

Introduction

A quick glance at a world map shows that if the continents were cut out like the pieces of a giant jigsaw puzzle, many of them could be roughly fitted together. In particular, the bulge of South America fits snugly into the hollow of Africa. As long ago as the year 1620, the famous scientist Francis Bacon commented that this could hardly be mere coincidence. Not long afterwards (1668) R. P. F. Placet, in France, wrote a book in which he claimed that the Old and New Worlds had once been joined, but had become separated at the time of the biblical Flood. In the seventeenth century, of course, the great age of the Earth was not known, and it was widely believed that it could be no more than a few thousand years.

The Flood was still in the thoughts of the German explorer Alexander von Humboldt, who in 1800 proposed that the Atlantic was nothing more nor less than a huge river which had broadened as a result of the torrent of water over which Noah had sailed in his Ark. Much later, in the 1870s, came a more rational theory by Sir George

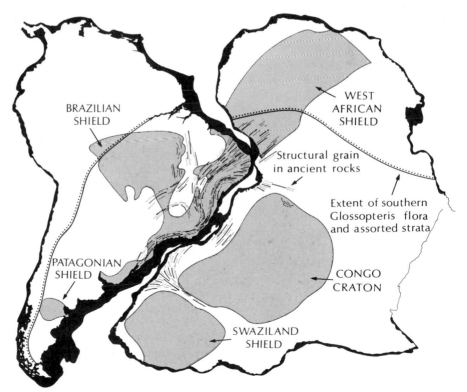

The 'fit' between western Africa and eastern South America. This is best achieved along the margins of the continental shelves. Note how geological features on one continent have counterparts on the other.

97

Darwin (son of Charles Darwin) which involved the Moon. According to Darwin, the Moon had once been part of the Earth, and had been thrown off as a result of the rapid rotation of the combined mass. This idea was extended, and it was suggested that the hollow now occupied by the Pacific marked the site of the Moon's departure.

Plausible though it may sound, the idea is completely untenable, if only because the Pacific depression is very slight compared with the bulk of the Moon. It was not until the early twentieth century that the idea of drifting continents was revived. There were several preliminary suggestions, but the real founder of modern theory was a German, Alfred Wegener. His first book on the subject was published in 1912. Wegener was not primarily a geologist: he was a meteorologist and also an explorer (he died in 1930 during an expedition to Greenland). His book was widely discussed, but few geologists took it seriously, and many pointed out errors in both detail and interpretation.

There were indeed discrepancies, but it was the principle that mattered. For instance, Wegener pointed out that there were gross similarities between ancient fossil remains found in eastern South America and western Africa: just where the two continents might be fitted together in the jigsaw. There were other geological similarities also, too many to be put down to sheer chance. This left only two possibilities: there could have been a landmass between the two continents, but there were very many reasons for rejecting that idea; or Africa and South America could once have been joined together.

For several decades Wegener's theories were almost forgotten in Europe, but in the late 1950s and 1960s there came a rapid and decisive change in outlook. The Australian geologist Warren Carey plotted the continents on a globe and made transparent outlines of them, so that they could be slid around on his globe; many fitted together. In 1965 Sir Edward Bullard employed a computer to do a similar job and confirmed Carey's results. Conclusive support came from palaeomagnetism, as well as isotopic dating and stratigraphic palaeontology. By 1970 the battle was over. Continental drift was no longer theory; it was an accepted fact. Forty years after his death, Alfred Wegener had been well and truly vindicated.

Continental Drift

One of the reasons why most geologists were initially so sceptical about Wegener's ideas was that there seemed no obvious reason why the continents should shift around or, indeed, how they could do so. The answer to this problem was given first by the great British geologist, Arthur Holmes, who suggested that because the solid mantle was at very high temperatures and under substantial pressure, it could actually flow slowly over long periods. Slow movements of this kind would be more than sufficient to drag along rafts of less dense lithosphere; although slow, the 'currents' would be extremely powerful. The situation is more complex than this, but Holmes certainly was on the right track.

To recapitulate: the Earth's surface is divided into a number of rigid plates, of which seven are large. The boundaries of the plates are marked by seismically and volcanically active zones. Continental drift really means the drift of the individual plates that carry the continents. Drift occurs because the plates are moving relative to each other, and the continents are carried along as part of the lithospheric plates. Once this basic principle has been accepted, the evidence is extremely

strong. For instance, the geological structures on the sides of continents that once were joined often fit perfectly together. Consider, for
example, the structure of the Saharan Shield, which is around 2000
million years old. The structural grain of the rocks runs north-south
toward the interior, but then swings roughly west-east toward the
Atlantic. There is a well-defined boundary between these ancient
rocks and younger rocks which runs into the ocean off the coast of
Ghana. If continental drift is valid, similar features should be found on
he eastern coast of South America. Indeed, this is so: the boundary and
a similar structural grain are found in Brazil.

Further support for the theory comes from fossils. Ancient fossils
retrieved from Africa and Greenland, for instance, show that during
the Silurian, Greenland was in the tropics, while at the same time
Africa was glaciated! Then comparison of certain index fossils found
in the Phanerozoic rocks of both Gondwanaland and Laurasia shows
that while they were at one stage separated by the wide Tethys ocean,
at other times they were very close together, if not actually joined. This
was particularly true between 350 and 220 million years ago.

The most convincing evidence probably comes from palaeomagnetism; we have already seen how the pattern of 'striping' matches
perfectly on either side of the Mid-Atlantic Ridge. We can also define
the past positions of the continents with respect to the magnetic poles.
There is very clear evidence that they occupied very different latitudes
from their modern ones. 'Polar wandering curves' (really a complete
misnomer, since it is the continents that have strayed, not the poles)
have been used to trace the paths of the continents from the Precambrian to recent times. To mention one finding: the data shows that
before about 165 million years ago, the Atlantic Ocean did not exist;
Europe and North America were joined at this time – there can be no
other satisfactory explanation.

The theory of plate tectonics is important not only for purely scientific reasons, it aids in the search for important natural resources, like
oil and natural gas. Perhaps it may one day even lead to a successful
way of predicting earthquakes and major volcanic eruptions – who
knows? The verification of continental drift has been as important in
geology as Darwin's theory of natural selection has been to the science
of biology.

Continental Jigsaw

The task of attempting to reconstruct exactly how the continents fitted
together at the beginning of the Palaeozoic era is a very trying one.
Various workers have made attempts to do this and, naturally enough,
there are substantial differences between their results. There are,
however, a number of points about which there is little disagreement:
all agree that North America and Greenland should be put back alongside western Europe, and that Africa should be placed alongside South
America. The former completed jigsaw gives us Laurasia. Today this is
separated by the broad Alpine-Himalayan mobile zone from the remnants of Gondwanaland.

South of the Alpine chain and to the west sit the continents of South
America and Africa, while to the east are the ancient stable blocks of
Australia, Antarctica and peninsular India. As we have previously
intimated, most geologists would restore all these pieces of the jigsaw
puzzle into one supercontinent, Gondwanaland, which existed during
the Late Precambrian. Today widely separated pieces of this ancient

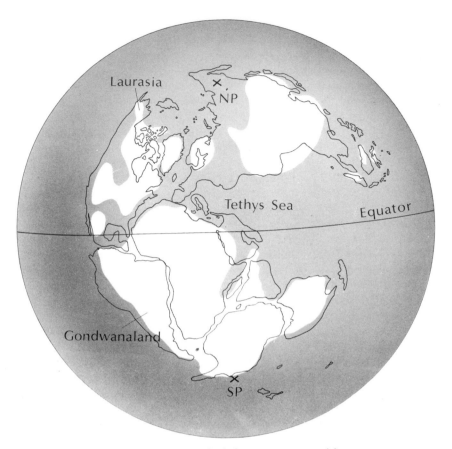

Continental jigsaw. One way in which the various pieces of the continental jigsaw fit together. Thus, in Triassic times, Laurasia and Gondwanaland became joined towards the west but were more widely separated to the east by the Tethys Ocean.

mass are found in the Mediterranean, the Middle East, the Himalayas and Central America.

Pangaea, in one form or another, seems to have existed in Late Palaeozoic times, some 200 million years ago, for Laurasia and Gondwanaland clearly were in contact in places, particularly along what is now northern Africa, the north coast of South America and the eastern coast of North America. To the east, however, there seems to have been a much wider separation of the two by Late Palaeozoic times, this great embayment being occupied by an early ocean, known to geologists as Tethys. This was to remain a major feature of the Earth's surface for many millions of years. Bearing in mind how the ancient continents were distributed in Late Precambrian and Early Palaeozoic times, it is clear that there were fewer 'small' oceans about, in fact it was almost a case of 'universal ocean' within which were scattered the ancestral continents.

By Early Palaeozoic times, Gondwanaland clearly existed as a single unit; however, Laurasia's pieces do not yet seem to have come together at this time. The evidence comes from palaeomagnetic data which shows that, during the Palaeozoic, the position of Siberia was very different from that of Europe; it was not until much later (in the Carboniferous and Permian) that they became welded together along a mobile belt called the Uralides.

The Ancient Oceans

Today's oceans are relatively recent geological features. We know that the oceanic crust that underlies them is nowhere older than about 200 million years. Plate tectonics theory explains how the relatively dense oceanic crust is continuously recycled by the Earth's heat engine, being generated from the mantle at mid-oceanic ridges and returned to the interior down subduction zones. Is there any way, therefore, in which we can learn anything about the ancient oceans which long since have been destroyed by subduction?

The first question to consider is what kinds of rocks might we be looking for? The ocean floor comprises submarine basalts, intrusive rocks of comparable chemistry, and an assortment of oceanic deep-water sediments, among which are shales, siliceous oozes and fine-grained volcanic muds. Some of the oceanic sedimentary and volcanic rocks which originally accumulated on the flanks of a ridge may have been swept down toward the ocean floor by dense turbidity currents and spread out there among the more typical deep-water sediments. Such rocks are called turbidites.

Given the situation where a slab of oceanic lithosphere is plunging beneath a continental margin then substantial slices of the oceanic rocks may be scraped off the plunging slab and may eventually thrust onto the margin of the continent. Naturally enough, rocks of this kind that have suffered such a fate will tend to be very altered and strongly deformed. Indeed will they be unrecognizable?

There is every reason to suppose that they ought to be, yet in recent times geologists have begun to suspect that they can track down such rocks, and in so doing, trace out the sutures that locate the old convergent plate margins. The particular kinds of rocks that have led to this optimism are called ophiolites and they have been discovered among the deformed strata of fold-mountain chains like the Appalachians and Urals, where it is suspected (e.g. from palaeomagnetic evidence) that ancient plates once collided. A typical ophiolite 'suite' contains deep-sea sediments, particularly shales, limestones, silica-rich rocks called cherts, and rocks redistributed by turbidity currents. Along with these are found submarine basalts and igneous rocks like gabbro and peridotite (although in a rather altered condition). The current view is that the ophiolites represent slices of oceanic crust together with upper mantle rocks, which were thrust onto land, where an ancient ocean disappeared.

Plate Movements and Orogenies

Although plate tectonics is now accepted by most geologists as a means of explaining the major changes that have affected the Earth's lithosphere, many of the details are still being argued about. We can be pretty sure that mountain building (orogeny) primarily occurs where lithospheric plates converge. In such situations, sedimentary deposits that have accumulated near to a continent's margin are crumpled up, the rocks being folded and thrust over each other. Accompanying such tectonic events is the rise of magma into the crust in the vicinity of the collision zone. The actual details may vary from place to place.

In some situations a plate with continental crust on its leading edge

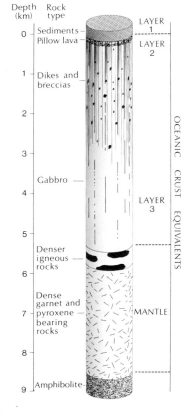

Section through a typical ophiolite sequence, showing the relationship of the various rock types encountered with normal oceanic crust and mantle as described in Chapter 8.

will be in contact with another carrying a continent, but some distance from the collision zone. Initially the oceanic lithosphere on the leading edge of the latter plate will be subducted beneath the former and fold mountains will be formed at the edge of the overriding plate. (A similar situation is found along the Pacific coast of South America.) Later, if subduction continues, a point may be reached when the leading edges of both plates carry continental crust. In this situation, because sial is too buoyant to be subducted into the mantle regions, plate movements would either be slowed down or actually stopped completely, if only for a while. Subduction might then start anew further away from the original collision zone, a new suture being formed and the whole cycle developing again.

A deposit commonly found along collision zones of the type described above is what is called a *mélange*. This is in effect a huge submarine landslide, instigated by the tectonic instabilities experienced at the margins of the colliding plates. Huge slabs of sedimentary and volcanic rocks may be literally scraped off the descending slab, to slither down the continental slope as a complex mass of coarse and fine debris. Such *mélange* zones are extremely complex, and the complexity is compounded by the metamorphism they suffer as they descend quickly downward into hotter regions, along with the downward-moving slab of cold lithosphere. Large-scale *mélange* can be found amid the Lower Palaeozoic rocks of western Newfoundland.

Mélange belts are accompanied by zones of intense magmatic activity; these usually run parallel to the *mélange*, and presumably arise as friction, generated during subduction, causes melting of the upper part of the downward-plunging slab as well as the water-laden sedimentary

Diagram to show the possible configuration of a convergent plate boundary where an island arc has formed. Many geologists believe that during their formation, island arcs separate into two zones: a frontal arc (a) along which there are both deformed oceanic sediments and active volcanicity, and a remnant arc (b) which lies closer to the continental margin and is not volcanically active. The two arc zones become separated by a marginal basin (c), which develops as lithospheric motion beneath it gives rise to rifting of the continental crust on which it is sited. On the landward side of the oceanic trench is what is called an accretionary prism (d) formed from oceanic sediments that have been scoured from the descending oceanic plate (e).

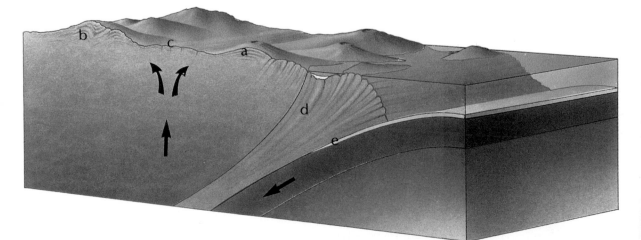

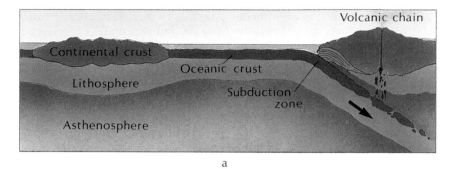

a

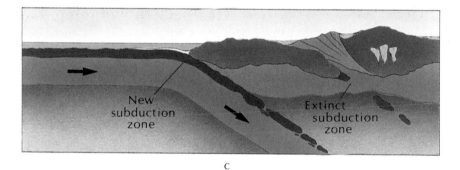

b

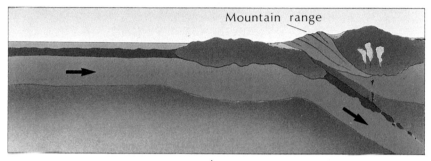

c

ABOVE Collision between two lithospheric plates. (a) Two plates, one
bearing oceanic crust, the other with continental crust at its leading
margin, converge. Magmatic activity and deformation of trench sediments
are characteristic of the overriding continental plate; so also are *mélange*
deposits. (b) Collision between the two blocks of continental crust occurs,
generating a new mountain chain, volcanic and plutonic activity. (c) In
some situations, the advancing plate may become disrupted, plate motion
may stop and eventually a new subduction zone is formed elsewhere.

LEFT *Mélange* deposits, Lleyn Peninsula, North Wales.

rocks plastered onto it. Magmas generated in this way form at depths of
between 100 and 200 km. Because they are relatively buoyant, they
rise quickly to be extruded along chains of volcanoes situated upon the
leading margin of the plate. Typically the lavas produced have the
composition of andesite. Some magma will solidify at deeper levels,
remaining there as bodies of granite rock.

14

LATE PRECAMBRIAN TIMES

Introduction

Toward the end of Proterozoic time the Earth experienced a prolonged period of crustal instability, during which a worldwide system of mobile belts developed. Most of these remained active until well after the beginning of the Palaeozoic era. In Laurasia mobile zones formed around a small number of small cratons. Many were to restabilize before the close of the Lower Palaeozoic. Activity within this system of belts spans the time period between 900 and 400 million years ago.

The Caledonian Mobile Belt

The term 'Caledonian' was first introduced by the Austrian geologist Eduard Suess (1831-1914), who applied it to a belt of strongly folded rocks found in Great Britain and Scandinavia, and overlain by deposits of Old Red Sandstone type (Devonian age). The cycle of events that produced these rocks culminated in the Caledonian orogeny, the rocks and the orogeny also being known as the Caledonides; its effects were felt well beyond the confines of the European region.

The earliest Caledonian events to be recognized date from about 800 million years ago. The first phase saw the formation of broad basins in which sediments began to accumulate. Before the close of the cycle these had an aggregate thickness of at least 15 km.

The active zone flanked the western edges of both the Baltic Shield and European Craton. Both of these ancient blocks still remain intact. However, on its west it flanked an ancient craton that has been fragmented. A tiny remnant of this is to be found in northwest Scotland, but the largest pieces are in central and west Greenland.

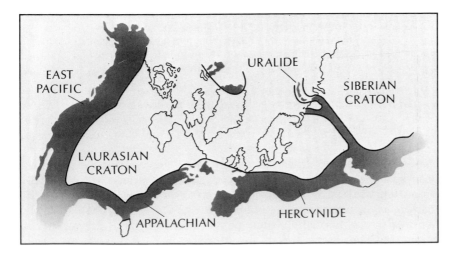

The mobile belts that formed along the margins of Laurasia in Mid-Palaeozoic times.

The Cycle of Events

The Caledonian cycle began with the formation of a number of depositional basins – downwarps in older cratonic blocks. Some of these formed early in the cycle, while others evolved much later. Orogenic activity began in the early basins even before sedimentation had ceased and continued well into Palaeozoic time, producing metamorphism and deformation of the accumulated sediments. Those rocks that collected in the early basins suffered quite intense changes, technically called high-grade metamorphism, but those of the younger basins escaped more lightly and were affected less severely; their metamorphism was of a lower grade.

About 400 million years ago the final phase of the orogeny began: the whole region was gradually uplifted to form a majestic new fold-mountain chain: the Caledonian Mountains. These would have stood proud for millions of years but eventually were worn down by erosion and the resultant debris spread out at the foot of the new mountain slopes or in new basins formed between adjacent massifs. These Old Red Sandstones were spread out on continental crust and differed markedly from the marine rocks of the mobile zone.

Later compression of the belt led to much fracturing of the rocks, which were also invaded by granitic magma. By about the middle of Devonian times (350 million years ago), the mobile belt had ceased to be an active region, and was certainly stabilized in northern Europe and the North Atlantic region, but stabilization was delayed in the more southerly regions of Europe and in North America. In these latter regions the mobile belt was transformed into a great fold-mountain chain during Late Palaeozoic times.

Sedimentary Basins

At the beginning of the Caledonian cycle sediments were laid down on an eroded land surface built from gneisses of Lewisian age. This represented a small fragment of an old craton that existed before continental drift, and the greater part of which now is found in Greenland. The earliest deposits, known to geologists as Moinian rocks, are a thick series of sandstones that appear to have built out into the sea as large deltas. They reached a thickness of around 7 km, and have been radiometrically dated at between 1000 and 800 million years old.

On the western side of the belt, where the sediments were in direct contact with the cratonic basement, there is what is termed a foreland region. It was here that the deformed sediments were thrust out over the cratonic basement later in the cycle. Here the sedimentary rocks are thinner, consisting primarily of shallow-water sandstones, boulders and pebble beds. These belong to what is called the Torridonian sequence which has been dated as being between 1000 and 800 million years in age. Later, 'Dalradian' rocks accumulated. These were predominantly deeper-water shales, limestones and immature sandstones, but in the upper part of the Dalradian succession are glacial horizons. The Dalradian sediments reached a similar thickness to the Moinian strata.

In east Greenland and in Spitsbergen there are similarly thick sequences of sediments which include sandstones, mudstones, limestones and tillites. In the former locality they occupy a broad zone, 200 km wide, containing rocks ranging in age from Late Proterozoic to Ordovician. In its core, the rocks have suffered high-grade metamorphism. In Scandinavia, the older basins immediately adjacent to the ancient cratonic foreland also collected sediments which range in age from Precambrian to Ordovician. Occupying a position further from the front are younger basin-filling rocks which include feldspathic sandstones and tillites, These may be up to 6 km thick and appear to be shallow-water deposits. During the later part of the cycle there appear to have been basins only in the southeastern part of Britain, where a series of subsiding elongated troughs received a substantial thickness of debris that included volcanic rocks and turbidites which are typical of tectonically unstable environments.

One of the characteristics of these Caledonian basin-filling deposits is the occurrence of glacial rocks called tillites. These bouldery deposits usually are found interbedded with marine rocks and have been attributed to glaciation during the 'Varanger Ice Age', when they may have been deposited from drifting sea ice. The glacial beds seem to occur stratigraphically below the fossiliferous Cambrian strata and are presumably of much the same latest Protozoic age wherever they occur in the Caledonian belt.

More about Basins

The basins in which the enormous volume of sedimentary debris collected must have been very large features indeed although most of the sediments accumulated in relatively shallow water. This fact was stumbled upon first by James Hall in the mid-nineteenth century. Hall, an American geologist based in New York State, noted that the deformed strata of the Appalachians were much thicker than those of the same age found in the American Midwest. In the Appalachians the total thickness is of the order of 12,500 m, but on reaching the valley of the Mississippi, this has dwindled to around 1500 m. Hall suggested that the Appalachian strata had accumulated in a slowly subsiding, long-lived trough. Some years later another American, James Dana, proposed that such vast troughs should be called 'geosynclines'. This term remains in the literature but has suffered much misuse and, to avoid confusion, we prefer not to use it; instead we shall simply refer

Schematic diagram showing Caledonian basins.

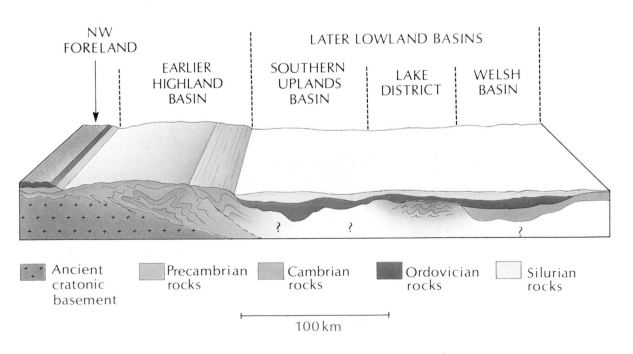

NW FORELAND

LATER LOWLAND BASINS

EARLIER HIGHLAND BASIN

SOUTHERN UPLANDS BASIN

LAKE DISTRICT

WELSH BASIN

Ancient cratonic basement

Precambrian rocks

Cambrian rocks

Ordovician rocks

Silurian rocks

100 km

ABOVE LEFT Photograph of Late Precambrian sedimentary rocks on the coast of eastern Greenland. In places the zone of sediments is 200 km wide and includes rocks that range in age from Late Precambrian to Ordovician.

ABOVE RIGHT Lower Cambrian turbidites at Hell's Mouth Bay, North Wales. The thicker beds are about 2 m thick. Each unit is the product of rapid deposition from a turbid sediment-laden submarine current called a turbidity current. These flowed down the flanks of the Caledonian basin during Early Palaeozoic times.

to depositional basins and identify specific features when necessary.

During Late Proterozoic and Early Palaeozoic times there appear to have been two main basin systems, each of which was composed of several smaller basins. The rocks now found in more northwesterly locations are very similar to the strata found in the Northern Appalachian basin, which is of similar age. Those towards the southeast on the other hand more closely resemble strata located in eastern Newfoundland, Nova Scotia and New Brunswick.

To take the Northwestern Basins first: their remains are to be found in the very northwest tip of Scotland and the western part of Spitsbergen, where limestones and other carbonate-rich rocks unconformably overlie Precambrian rocks. These apparently were situated at an ancient passive continental margin. Further to the southeast, however, are deformed rocks called schists and amphibolites, which were marine sediments and volcanic rocks before they became metamorphosed. These accumulated in deeper water and now underlie a belt that runs from northwest Ireland, through northwest Scotland and extreme western Norway into eastern Spitsbergen. Although the majority of these rocks can be shown to rest on sial, toward the southeastern margin of the belt there are indications that the basins lay on oceanic crust, on an active margin.

The southeastern basins extended from southern Ireland, through northwest England into western Norway and on toward Spitsbergen. The rocks are shales, greywackes, sandstones and a variety of basaltic and andesitic volcanic rocks; numerous unconformities occur within sequence, indicating periodic upheavals. These presumably accumulated along an active continental margin. The volcanism and deformation that occurred during Early Palaeozoic times may have been produced by passage of a plate bearing ancestral North America beneath that containing ancestral Europe at this time.

Continental Shelf Deposition

Sediments that are laid down under the relatively quiet conditions of a passive continental margin gradually give rise to orderly sequences in the large basins. In time, a wedge-shaped accumulation of sediment forms, this having been eroded from the continent itself. This forms part of a broad offshore shelf.

As new oceanic lithosphere cools and contracts after it has emerged from a spreading centre, so the trailing edge of the attached continent slowly subsides. In so doing it allows the basin to collect land-derived debris over a very long period; as a result the sedimentary burden depresses the crust still further, enabling the basin to collect more and more sediment. Calculations indicate that for each 2 m of crustal subsidence, about 3 m of sediment can accumulate. In this way accumulations may greatly exceed 10 km thickness.

Beneath such a typically wedge-shaped body of sediment it is usual to find rift valleys incised into the crust below. These would have opened in response to earlier tensional stresses within the continental crust. Basaltic lavas and continental deposits would have been built up on the floors of the rifts, much in the same way as they have in the East African Rift in recent times.

Initially, sandy materials begin to fill the basin. Much of this relatively coarse debris will be dumped on the continental slope and, during seismic activity (a characteristic of such environments), this may be shaken off and flood down the submarine slope as a turbid suspension of sand, mud and water, called a turbidity current. Rocks formed by this kind of activity (turbidites) become interbedded with the normal fine-grained mud rocks that accumulate in deeper waters. Very thick sequences may develop in this way.

Later in the life of a trough, as a shelf of sediment gradually builds out, so sedimentation may become dominated by shales and carbonate rocks, because the supply of sediment from the adjacent continent diminishes. This is a pattern seen many times in the stratigraphic record, and is all part of the geological cycle.

The Caledonian Forelands

When we look along the western margin of the Caledonian mobile belt, that is in northwest Britain and Greenland, we find that there are no strata younger than mid-Ordovician. In northern Greenland, however, the old craton is covered by a sequence of limestones and shales that are fossiliferous and span the time interval Cambrian to Late Silurian; similar deposits are found also in northern Canada. Their presence shows that shallow seas spread over large regions of the ancient craton during these times.

On the opposite side of the mobile belt, that is, in Britain and in Scandinavia, similar rocks are found. In Shropshire, England, Lower Palaeozoic shallow-water sandstones and other shoreline deposits bear witness to the presence of a shallow sea, out of which the well-known viewpoints of the Long Mynd and the Wrekin must have risen as islands. In Norway in the region of Oslo and in southern Sweden, shallow-water sandstones, limestones and shales of Lower Palaeozoic age also occur, indicating that similar conditions prevailed here, too. By Late Palaeozoic times, however, the foreland regions gradually emerged from beneath the waters as the latter stages of Caledonian activity were reached.

In the core of the mobile belt tremendous upheavals rucked up the

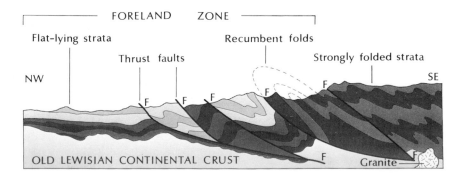

Schematic section through the foreland region of the Caledonian mobile belt. Flat-lying shelf sediments typify the region northwest of the main thrusting. To the east these may be involved in the thrust faulting. Further east again, large recumbent folds typify the Moinian and Dalradian rocks.

BELOW Folded and faulted Cambrian strata, Porth Ceiriad, North Wales. The thick-bedded sandstones (right) have been faulted against thinner-bedded shales and siltstones (left).

BOTTOM Minor folds in Late Precambrian rocks at Uwchmynydd, North Wales. The vertical lineations running through the strata are cleavage planes, a tectonic structure impressed upon the rocks during Caledonian orogeny.

sedimentary strata into large folds, many of which became 'recumbent': that is, they were forced over so as to rest horizontally. Tremendous tectonic forces drove many of these folds outward from the core of the orogenic zone and they were thrust over the ancient cratonic basement gneiss and also metamorphosed sediments from within the core regions are now found as thrust sheets between low-angle fractures called thrust faults. These are characteristic features of foreland regions. One of the best-known examples of such a fault is the Moine Thrust of northwest Scotland, a major zone of lateral displacement where movements of several tens of kilometres were experienced by the rocks involved in these terminal movements. Radiometric dating of the rocks altered by these events indicates that the last movements along the Moine Thrust occurred about 430 million years ago.

Folding and Metamorphism

One of the characteristic features of sedimentary and, indeed, many volcanic rocks, is that they are more or less horizontal when laid down. A brief look at the rocks of any orogenic belt shows that in these unstable regions of the Earth's crust horizontal beds are the exception. In fact they have been rucked into folds that vary in size from structures tens of kilometres across to tiny microfolds measurable in millimetres. Folding is a part of the important geological process of deformation or tectonism.

Not every rock responds to tectonic force in the same way; thus, a brittle sandstone takes up the strain in a very different way from, say, a shale or mudstone. The exact manner in which rocks deform is dependent upon properties such as their chemistry, the amount of water they contain in the pore spaces between their constituent grains, and upon such external factors as the pressure and temperature under which they are confined and the rate at which they are compressed at any given time. The same rock, confined under different conditions, will respond in quite different ways to the stresses imposed upon it. As an illustration consider a slab of toffee, which will break under the hammer but which flows slowly under prolonged stress.

Whatever the exact situation, the outcome of deformation is usually the production of folds in rocks, or, if the rocks behave in a rather more brittle way, of fractures or 'faults'. Perhaps the San Andreas Fault is the world's best-known fault, but there are innumerable such structures developed in the Earth's crustal rocks, countless millions of which are far less imposing than their illustrious Californian superior. Folds and faults both have varying styles. In the previous section the term recumbent fold was mentioned. Such a fold has an attitude whereby the axial plane of the fold is near horizontal. In other kinds of folding, the axial plane may be vertical or at some other intermediate angle.

Every period of deformation produces its own particular fold regime – the Caledonian orogeny was no exception. During the long span of time during which the Caledonides were operative, the rocks within the mobile zone were deformed many times, as many as five or six times in some instances. As a result folds, refolded folds and refolded refolded folds are seen – the situation can become extremely complex! One of the tasks of the structural geologist is to unravel the folding complexities of deformed rocks such as those of the Caledonides.

Deformational processes take place at considerable depth inside the Earth. As we have already mentioned, both the temperature and pressure gradually rise towards the centre of the Earth, so that where folding

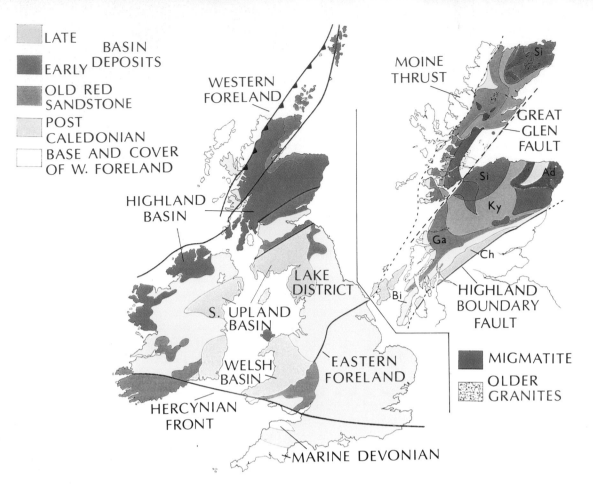

WESTERN FORELAND

MOINE THRUST

GREAT GLEN FAULT

HIGHLAND BASIN

Si

Si

Ad

Ky

LAKE DISTRICT

Ga

Ch

S. UPLAND BASIN

Bi

HIGHLAND BOUNDARY FAULT

WELSH BASIN

EASTERN FORELAND

HERCYNIAN FRONT

MARINE DEVONIAN

is occurring at a depth of, say, 10 km, the rocks are under high pressure and experiencing very elevated temperatures. Since sedimentary rocks are actually generated at surface temperatures and pressures, it is not surprising to learn that they feel decidedly uncomfortable in their new environment! In fact the original constituents of the rocks will rearrange themselves, forming either larger crystals of existing minerals, or new minerals that are stable under the new conditions. Such change, termed metamorphism, takes place on the atomic scale, in the solid state, and does not involve actual melting of the rocks.

One of the first geologists to study the effects of metamorphism within an orogenic belt was George Barrow who during the latter part of the nineteenth century studied the Dalradian rocks of the British Caledonides. He noted that if you crossed the Grampian Mountains from southwest to northeast, a succession of different assemblages of 'index' minerals appeared in the rocks. These minerals, produced in response to regional metamorphism of the Dalradian sedimentary rocks during the Caledonian orogeny, apparently defined zones of successively changing grades of metamorphism, the most intensely changed rocks occurring toward the northeast.

Barrow, in mapping out these zones, had stumbled upon the fact that during metamorphism a thermal gradient is established. In moving westward away from the 'core' regions, you eventually reach rocks of very low grade which lie adjacent to the foreland zone. A similar traverse toward the north and south also takes you into less strongly metamorphosed rocks. In fact, in any orogenic belt, the most highly deformed and metamorphosed rocks are those that were closest to the core of the mountain belt, or 'orogen' as it is technically named.

The Caledonian mobile belt as developed in Britain. (LEFT) the main structural elements of the belt; (RIGHT) an enlarged map of the Highland Zone showing the arrangement of Index Minerals produced in the Caledonian rocks during metamorphism. Ch=chlorite; Bi=biotite; Ga=garnet; Ky=kyanite; Si=sillimanite; Ad=andalusite.

ABOVE LEFT Banded gneiss cut by granitic vein. The banding was produced by the segregation of light and dark silicate minerals while the rock was deeply buried.

ABOVE RIGHT Granite invading bedded sedimentary rocks, Cruachan, Scotland. Note how the light-coloured igneous rock has pervaded the weaknesses between the bedding planes in the invaded sediments.

RIGHT Caledonian granite, showing prominent pink feldspar crystals, together with black mica and greyish quartz.

Caledonian Magmatism

Volcanic activity accompanied the growth of the Caledonides. Much of this was submarine in nature – witness the pillow lavas found in Wales and southwest Scotland, but there were significant volcanic centres also on the continental margins and these may have represented island arcs. Basaltic lavas are very abundant within the marine sedimentary succession, while andesitic and rhyolitic lavas are common, particularly within the Ordovician strata of Wales and the English Lake District.

Perhaps the most important magmatic event was, however, the rise of vast quantities of granite magma into the crust during the latter stages of the cycle. There are at least fifty separate bodies of granitic rocks in Britain alone; about one tenth of these predate 450 million years ago, the rest are younger. Similar numbers occur in Scandinavia, Greenland and Spitsbergen. Radiometric dating indicates that although the range of ages is between 600 and 390 million years, a significantly large proportion were emplaced around 400 million years ago, and this is perhaps the last major thermal event of the orogenic cycle.

Not all of the granite bodies were formed in the same way. Some show evidence of having forcefully pushed their way up into the crust. Others have a complex emplacement history, such as the one in Glencoe, Scotland, which features widely in the geological literature. In the central regions of the British Caledonides the granites are associated with migmatites: rocks that have a mixed magmatic-metamorphic parentage, and are characteristic of the root regions of orogenic zones. Elsewhere granite bodies have metamorphosed the rocks of the surrounding country, indicating that the latter were relatively cool when the magmas rose into them.

Accompanying the major bodies of granite are pegmatites, occurring as swarms of smaller igneous bodies – dikes and sills. These are sheet-like in form and in many instances bear witness to the activity of volatiles which escaped from the larger igneous masses during the later stages of their crystallization.

The Final Stages

Toward the end of Silurian times, that is, about 395 million years ago, a number of important changes in the geological situation had occurred in Europe and the North Atlantic region. Firstly, temperatures within the crust had fallen perceptibly within the mobile zones. This is shown by the abundance of K-Ar radiometric ages around 410-450 million years – the rocks then being cool enough to allow retention of argon to begin, thus 'setting the radiometric clock'. Secondly, most of the folding appears to have ceased and, thirdly, changes occurred in the pattern of sedimentation, new basins of deposition being formed on continental crust. These basins were to receive a great thickness of sediments that were formed on land, from the newly rising Caledonian mountains. Add to this the rise of enormous quantities of granitic magma between 410 and 380 million years ago, and we see a great deal of activity crammed into a relatively short span of geological time: about ten per cent of the cycle's full length.

This, then, was the close of the Caledonian cycle. It was marked by the rapid erosion of the new mountain chain which rose isostatically to compensate for its greatly thickened crust. This resulted in the accumulation of thick continental deposits traditionally known as the

Old Red Sandstone. As we might expect, these deposits were thickest toward the margins of the old mobile belt, for these marked the lower flanks of the new fold mountains. Over 8 km of such strata were laid down in east Greenland at this time. The Old Red Sandstone, however, is part of another story.

Precambrian Ice Ages

The Caledonian cycle spanned the Late Proterozoic-Early Palaeozoic period of time. It provides a link between the very long period where geological data are gleaned with extreme difficulty, and not without some measure of uncertainty, from more recent epochs when fossilizable life had clearly evolved and from whose rocks a more certain history can be recovered. Despite the difficulties associated with Precambrian stratigraphy there is more than circumstantial evidence to support the notion that glaciations occurred early in Earth's story.

There does not appear to be any evidence for even one ice age during Archaean times. In part this might be due to the absence of very large masses of continental crust for ice sheets to get established on; or it might be partly explained by unsuitable climatic conditions. Whatever the case, evidence is lacking.

The situation is rather different for the Proterozoic, however, and there are strong grounds for supposing that widespread ice sheets covered parts of the North American craton several times. Furthermore, supposed Proterozoic glacial deposits have been reported from all of the continents except South America. The kinds of rocks found that strongly suggest glaciation are boulder-bearing tillites, rhythmically banded sediments akin to the 'varved' deposits of the most recent ice age, and 'dropstones' – boulders that appear to have been dropped from floating ice into submarine sediments.

Some of the best evidence comes from the Early Proterozoic rocks of the Lake Huron region of Canada, where among 12,000 m of sedimentary rocks are three probable glacial horizons. The youngest of these, called the Gowganda Formation, is very extensive, covering an area at least as large as 120,000 km^2. It is considered to be about 2300 million years old. Similar rocks are found elsewhere in Canada, and detailed study of the rocks in this region suggests that during this period, sediment was being deposited in a radial pattern, as though laid down around the periphery of a continental ice sheet.

On other continents a search for glacial deposits of similar age has been made, and there is good evidence among the Griqualand Tillites of southeast Africa, and the Turee Creek Formation of northwest Australia, for glaciation at roughly the same time. Palaeomagnetic evidence, although rather scanty, indicates that the centre of glaciation was probably around a southern palaeolatitude of 60 degrees from the Proterozoic Equator.

In the latter part of Late Proterozoic time, particularly between 950 and 570 million years ago, there appears to have been widespread glaciation: the first between 950 and 900 million, the second around 750 million, and the third around 600 million years ago, just before Cambrian time. Glacial deposits related to these events have been located on all of the continents save South America.

Exactly why glaciation occurred at this time is a matter of much debate. Some believe it to be due to the drifting of the early continents through the polar regions during the first really widespread episode of continental drift. Others would cite a rather special type of mountain-

Late-Precambrian varved sediments, Ella Island, Greenland. The banding is the result of the glacier's seasonal cycle of freeze-up and melt which releases pulses of sediment into lakes or the sea during the warm season. The coarser sandy bands were deposited quite rapidly but the intervening silty bands and clays took much longer and are thus much thinner.

building associated with drifting, while others again see an explanation in an 'anti-greenhouse' effect, whereupon carbon dioxide was lost from the atmosphere just before glaciation. It is probably correct to say that at present no one explanation is completely satisfactory. We must await further research before a more definite answer may emerge.

Blueprint for a Geological Cycle

The geological processes which accompanied the development of the Caledonides were neither confined to that particular period of time nor affected only the regions described. Similar processes have affected the crust and lithosphere elsewhere, at both the same and at different times. We have chosen to introduce the Caledonian mobile belt simply as it has been widely studied and provides us with a 'blueprint' for other geological cycles.

Continent-sized portions of the Earth's crust whose rocks, fossils and geophysical evidence imply that once they were situated at quite a different latitude clearly must have moved since they formed. Similarities between blocks now widely separated, perhaps by ocean, perhaps by belts of younger rocks, allow us to piece together the fragmentary evidence and show how continental collisions and splittings apart have shaped and reshaped the past or *palaeo*-geography of the Earth. Let us now look at other mobile belts and see how their study allows us to extend our palaeogeographical knowledge.

15

THE APPALACHIAN STORY

North America in Proterozoic times was very different from the continent we know today. As we have already shown, the ancestral continent did exist and at times formed part of a larger supercontinent, Laurasia. During the latter part of Proterozoic times the cratonic cores of North America, Europe and Gondwanaland were in fact separated; indeed, they were drifting apart. Later, in the Early Palaeozoic, thick deposits of sedimentary and volcanic rocks accumulated in elongate basins which developed along the continental margin of ancient North America. The plate movements apparently were then reversed and eventually the sediment-filled basins of the continental margin were

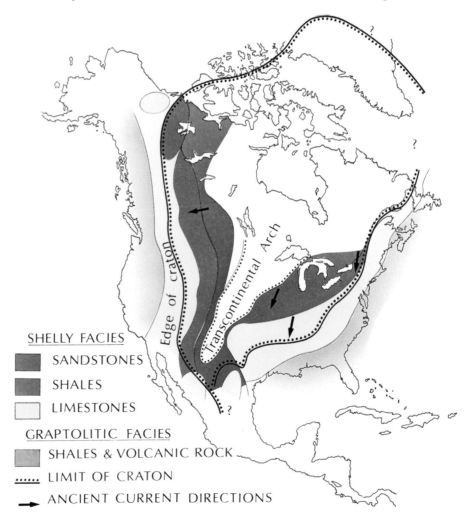

Ancient North America during the Late Cambrian. The margins of the craton at this time are indicated by the dark-black line. The main belts of deposition lay at the edge of the continent, where there was an outward passage from shelly to graptolitic facies. The presence of sandstones, shales and some limestones within the craton edge line reflects how the seas transgressed over the lower-lying regions of the continent in Late Cambrian times. The Transcontinental Arch apparently separated two regions of shallow-water deposition and began to rise during the Early Cambrian.

SHELLY FACIES
- SANDSTONES
- SHALES
- LIMESTONES

GRAPTOLITIC FACIES
- SHALES & VOLCANIC ROCK
- ..... LIMIT OF CRATON
- ➤ ANCIENT CURRENT DIRECTIONS

deformed to form part of the worldwide system of orogenic belts of which the Caledonides was also a part. Those that developed along the margins of ancestral North America are called the Appalachian – the major ones being Ouachita, Cordilleran, Franklinian and East Greenland mobile belts.

The Continental Interior

Over 5000 m of Early Palaeozoic marine sedimentary rocks are found along the eastern margins of North America. Calculations indicate that in the strip between Newfoundland and north Alabama, the deposits accumulated at the rate of about 13 m per million years. In contrast, in the continental interior, a maximum of only 1500 m is found, and during the Proterozoic, hardly any marine sediments appear to have been laid down at all. This implies that the whole of the craton was very stable. This in turn probably implies that the whole of the continent formed a low-lying land region.

As we move toward what were the old mobile zones, however, so the thickness of Late Precambrian sedimentary deposits increases; in places the oldest Cambrian strata lie conformably on the uppermost Precambrian beds. The facies variation shown by the sediments, together with the current directions revealed by structures within individual beds, indicate that the trough sediments had their source in the continental interior.

Evidently the ancient cratonic surface had been ravaged by weathering and eroded for at least 500 million years before the onset of major sedimentation in Late Precambrian times. Because of the longevity of these events, the deposits became 'mature' as less resilient minerals

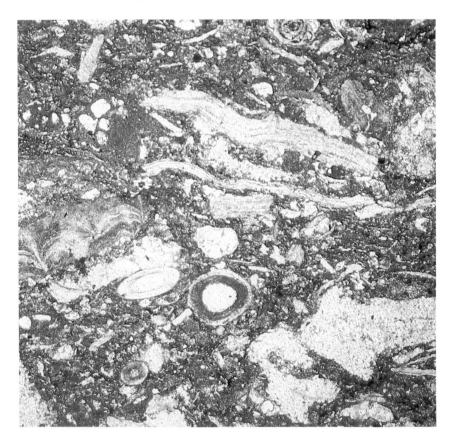

Photomicrograph of a shelly limestone, showing shell fragments set in calcite mud. The larger shells are about 4 mm long.

were either dissolved or broken up and dispersed, leaving only re-sistant ones like quartz. These were smoothed, sifted and sorted and then swept toward the continental perimeter by rivers and streams, there to build the relatively thick beds of 'clastic' sediments found at these margins. Among the Late Proterozoic strata are tillites of com-parable age to those found at the same point in the sequence in Eurasia: clearly glaciation affected North America, too.

A little later, during Early Cambrian times, a broad crustal arch warped the craton, extending from Arizona into the region of the present Great Lakes. This appears to have separated regions on either side over which marine deposition took place, as a shallow sea ex-tended over the craton during Cambrian times, covering vast areas by the end of that period. This Cambrian 'marine transgression' recorded by unusually widespread sedimentary deposits was one of the most widespread geological events of that time.

The Cambrian rocks are rich in fossil remains, including trilobites, brachiopods and cephalopods: all creatures that lived only in sea water. There is little doubt, therefore, that the cratonic cover is of marine origin. In all likelihood, the craton rose up and down at various times (this is called epeirogenic movement) throughout this period, so that positive regions were separated from one another by depositional basins, the positions of which altered as time passed. The arched regions would have supplied clastic materials to the Cambrian seas.

By Late Cambrian times, almost all the land seems to have gone; the source of the clastic sediments thus was submerged. Above the domi-nantly clastic rocks, therefore, we find the succession dominated by carbonates – like those currently being precipitated in the vicinity of the Bahamas Bank. Most of the rocks are composed of the broken shells of dead calcareous organisms: the kind of animals that thrive in well-oxygenated, agitated, shallow sea water. Such conditions are found today most widely in the tropics and subtropics; a fact that strongly suggests much of North America experienced a very equable and tropical climate at this time. This is supported by palaeomagnetic data.

The Appalachian Mobile Belt

The Appalachian Mobile Belt developed along the junction between ancient North America, Europe and Gondwanaland. It represented the site of several elongate marginal basins which accumulated a variety of rocks, before these continents converged in the Early Palaeozoic. When the intervening ocean was subducted and they eventually col-lided, as the lithospheric plates carrying the ancient continents and their marginal accumulations of sedimentary debris came together, the mobile zone was crushed and a new fold-mountain chain emerged where the ocean had been. Today, the eroded remains of these great ranges are found in a belt of deformed and metamorphosed rocks that extends from Newfoundland to Alabama.

Thick accumulations of sedimentary rocks typify the belt. At the beginning of Cambrian times, marine sediments accumulated only at the continental margins but moved westward gradually during the Early Cambrian, covering the shallower basins, too. Little deformation appears to have occurred until the middle of Ordovician times, when substantial upheavals took place. These were the Taconic Orogeny. In western Newfoundland, sheets of oceanic lithosphere – now forming ophiolite – were thrust towards the continental interior. These events have been dated at around 510 million years ago. In all probability a

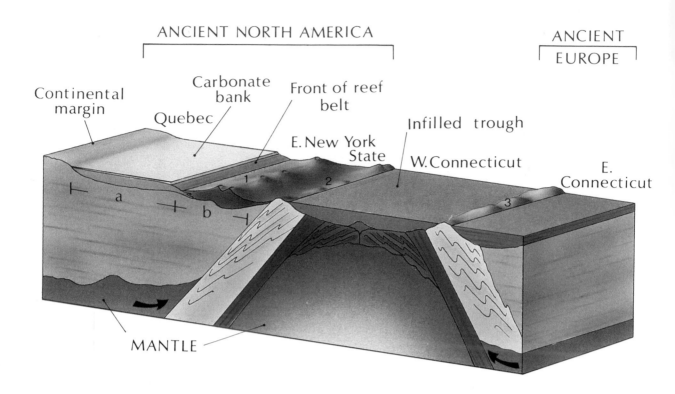

ANCIENT NORTH AMERICA

ANCIENT EUROPE

Continental margin

Carbonate bank

Quebec

Front of reef belt

E. New York State

Infilled trough

W. Connecticut

E. Connecticut

MANTLE

One possible reconstruction of the geology of the Northern Appalachian mobile belt during Cambrian times. The infilled trough, now underlying western Connecticut, was formerly the site of the Iapetus Ocean. 1=inner volcanic arc; 2 and 3=outer tectonic arcs; a=Inner Appalachian basin; b=Outer Appalachian basin.

subduction zone sloped westward beneath the Appalachian Belt. At roughly the same time, massive slides occurred, giving rise to extensive *mélange* deposits which have middle Ordovician strata beneath, within and above them.

In the Late Ordovician, the thick prism of sedimentary and volcanic rocks was intensely crumpled and fractured. In the southeast of New York State and in northern New Jersey, deformation was intense enough to transport slices of Precambrian gneisses, torn from their cratonic roots, bodily westward over the Ordovician strata, in much the same manner as Lewisian gneisses are seen above Palaeozoic rocks in the Moine Thrust of the Caledonian Belt.

Thus, these Taconic movements produced new fold mountains. The uplifted rocks were then rapidly eroded during Silurian times, as the sea attacked its borders. In central Maine, parts of Quebec and central Newfoundland there is evidence for a later phase of folding, metamorphism and intrusion during Late Silurian to Early Devonian times. Radiometric dates obtained for granites intruded into the roots of the northern Appalachians during this phase give an age of 395 million years for the event. This roughly corresponds to the time of the latest events of the Caledonian orogeny in Europe.

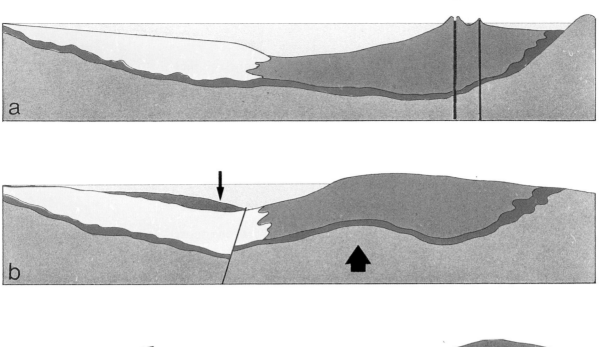

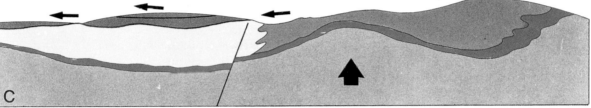

The production of sideways-moved slabs of crust (allochthons) during the Taconic movements in the region shown to the left of the previous figure. (a) Deposition of trough sediments during Cambrian times; carbonate-shelf deposits in the nearshore trough, deeper-water sediments further out; (b) uplift of the outer trough gives rise to deposition of clastic sediments on top of the carbonate sequence; (c) gravity-sliding of uplifted outer-trough sediments (allochthon) on to the inner-trough area.

Iapetus: Precursor of the Atlantic

As with all orogenic episodes, the Taconic upheavals that affected the northern Appalachians were directly related to lithospheric plate movements. Thus, in the latter part of Precambrian time, roughly between 800 and 700 million years ago, ancient North America split away from Gondwanaland – specifically that part of it now represented by Africa. This led to the opening of a new ocean called the Iapetus. Ancient spreading centres must have bisected it, just as they do in the modern Atlantic. For Iapetus was precursor to that great ocean. Iapetus widened for a long time but then, about 600 million years ago, during the final stages of Precambrian time, it began to close again. This brought about the convergence of plates during the Cambrian and eventually generated folding, faulting and volcanicity, which left their marks indelibly among the rocks of eastern North America.

It is thought that during the Late Precambrian a substantial fragment of continental crust became separated from the main American craton; a marginal sea developed over the oceanic lithosphere between them. On the receding margins of this and of 'mainland' America, shallow-water shelf sediments accumulated. In the Early Cambrian, the Iapetus ocean began to close; closure continued from Middle Cambrian times until the Early Silurian, a time interval of about 50 million years. During the early stages, there was both subduction adjacent to the plate margins and probably the development of an island arc or arcs. Subsequently the estranged continental fragment crashed into North America, with much thrusting and folding of rocks along the convergence zone. This produced the Taconic Orogeny in the northern part of the Appalachians. Eventually, between 500 and 400 million years ago, the island arc also was pushed into North America, producing a second phase of folding and thrusting which was even more intense than the first. This series of events, known as the Alleghenian Orogeny, rucked up the southern part of the Appalachian belt, as the more southerly parts of Iapetus also closed.

It is calculated that the lateral movements produced by the combined orogenies pushed some of the thrust sheets at least 250 km westwards over the margins of ancestral North America. Iapetus was eventually eliminated completely during Late Carboniferous or Early Permian times.

Present-day Features of the Appalachians

If we look at the present-day geological features of the southern Appalachians we perceive four distinct zones or provinces with their own particular characteristics. These are a direct result of their varying positions in relation to plate margins when the main orogenic episodes that fashioned the Appalachians occurred.

In the Valley and Ridge Province, thick Palaeozoic sedimentary rocks are found. These are intensely folded and show clear evidence of having been thrust northwestward by orogenic forces from the southeast. Three upheavals occurred, one at the close of Ordovician times, a second at the end of the Devonian, and a third in the Permian-- Carboniferous period.

Lying to the southeast is the Blue Ridge Province. The much-eroded mountains in this belt are constructed from Precambrian and Cambrian crystalline rocks that have been metamorphosed and deformed. They are, however, now separated from the Valley and Ridge rocks by a major dislocation: a thrust fault, which carried the gneisses northwestward from their original position and deposited them partly on top of the Palaeozoic sedimentary rocks.

Southeastward again lies the Piedmont province. Here Precambrian and Palaeozoic metamorphic rocks occur. These were originally volcanic and sedimentary rocks that were deeply buried, then invaded by granitic magma. Subsequently they were highly deformed, much eroded, then thrust northwestward over the Blue Ridge rocks. The volcanism began about 700 million years ago and continued, more or less unabated, for a further 200 million years. At least two episodes of deformation affected these rocks, one at the end of Devonian times and a second during the early part of the Carboniferous.

In contrast, the rocks of the Coastal Plain have been relatively undisturbed. The oldest date from the Jurassic period (about 200 million years ago) and are underlain by rocks similar to those now found in the Piedmont region. The present continental shelf is a submarine extension of this marginal zone. Recent geophysical work has revealed that the younger sedimentary rocks of the Valley and Ridge province continue eastward and actually underlie the older metamorphosed rocks of the Blue Ridge and Piedmont provinces. The

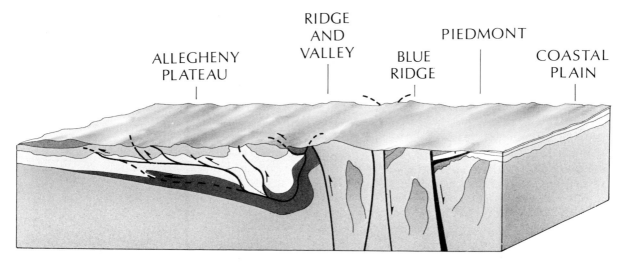

ALLEGHENY
PLATEAU

RIDGE
AND
VALLEY

BLUE
RIDGE

PIEDMONT

COASTAL
PLAIN

Present-day features of the
Appalachian region. The
vertical scale is, of course,
grossly exaggerated.

rocks of the latter two provinces were evidently metamorphosed before being thrust northwestward over the continental shelf sediments of ancient North America, which seems to have grown by the piling-up of thrust rock masses along its margins, at least during the life of the Appalachian mobile belt.

The Appalachians in the Late Palaeozoic

Ancient North America was more or less girdled by large sedimentary basins during this period. The Cordilleran orogenic belt ran down what is now the western margin and to the north lay the Franklinian belt. The Appalachian and Ouachita orogenic belts ran down the east and southeast sides of the old craton. Within the shallower inner basins and along the adjacent continental platforms, limestones and shales were laid down. In the deeper, more disturbed basins, coarser debris accumulated, particularly during orogenic episodes. The Late Palaeozoic was a time of tremendous upheaval in the Appalachians and before its close more fold mountains had been generated. At approximately the same time, parts of Eurasia were feeling the effects of the Hercynian orogeny.

The Alleghenian orogeny affected parts of ancient North America during Late Palaeozoic times and its effects were so intense that the Appalachian orogenic belt was again rucked up into a great mountain range, particularly in the region of the Southern Appalachians. The Ouachita Mountains of southwest Arkansas and southeast Oklahoma and the Marathon Mountains in north Texas were also generated at roughly the same time. All of these ranges have, of course, been deeply eroded since Palaeozoic times.

What was the cause of this great upheaval? When palaeomagnetic data, the radiometric ages of granites and metamorphic rocks in the various orogenic belts, and correlations between folded sedimentary strata are all taken into account, there is every reason to believe that all the continents were close together and converging on one another in Carboniferous and Permian times. It appears that the most likely cause of orogeny in the Appalachians at this time was the initial contact between the drifting continents of North America and Gondwanaland. It was not until the middle of the Permian Period that the two ancient continents seem to have stabilized and become completely welded together.

Landsat photograph of the Appalachians in central Pennsylvania. This superb picture of a classical fold belt covers the entire width of the valley and ridge province, a region characterized by alternating anticlines and synclines. To the west of this region (left-hand side of image) is the Allegheny Plateau, a region rising between 300 and 900 m above sea level and built from gently folded Palaeozoic sedimentary strata. Within the main Appalachian fold belt the intensity of deformation increases toward the southeast (note how the valleys become increasingly narrow in that direction). Next comes the Great Valley, eroded out of Palaeozoic limestones, while to the east again (bottom right), is the Blue Ridge Anticlinorium, a resilient block of ancient Precambrian igneous and metamorphic rocks.

The American Interior

During the Late Palaeozoic most of the craton was covered by a shallow sea. The more rapidly subsiding basins attracted substantial volumes of sedimentary deposits, which presumably were derived from hilly regions between them. There is certainly evidence for the uplift and increasing extent of highlands in the east and south of the craton during the Late Carboniferous, as increasingly large amounts of clastic debris were spread out in the eastern, central and southern parts of the continental interior.

The sea, however, did not remain at the same relative level throughout this time. There were repeated incursions and regressions over the almost flat continent as the sea level fluctuated, often over rather short time intervals. This gave rise to a distinctive kind of rhythmic sedimentation, in which lithologies such as shale, sandstone and limestone are repeated over and over again producing units known as cyclothems.

The sea finally withdrew from the eastern side of the craton in Permian times, whereupon coastal lagoons developed in what is now Oklahoma and Kansas. The large Delaware Basin formed among the hills of the Marathon Mountains of northern Texas and southeast New Mexico. The deeper parts of this basin became flanked by reefs in

The Grand Canyon, Arizona. This amazing declivity, worn by the Colorado River as it wends its way across the interior plateau of the continental interior, exposes a sequence of sedimentary rocks that ranges from Precambrian to Permian. In this particular view the canyon above the obvious terrace is built from shales, sandstones and limestones of Permian age which are separated by an unconformity (at the level of the ledge) from Pennsylvanian shales and sandstones. Nearly all of the beds are richly fossiliferous and provide geologists with an almost unprecedented record of life during this period.

16

LATE PALAEOZOIC MOBILE BELTS

Introduction

Caledonian tectonic events ended in Middle Devonian times, as did those affecting the northern Appalachians. At this time North America and ancient Europe were joined and probably lay between latitudes 20 degrees north and 30 degrees south: spanning the Equator, which is believed to have run through northern Norway, central Greenland and

The configuration of the continents during Devonian times.

across northern Canada into Alberta. A broad zone of rocks deformed in the Late Palaeozoic stretches across Europe and encompasses large areas in Spain, France and southern Britain. It also extends into North Africa. Similar deformed strata run the length of the Ural Mountains, are found in parts of Iran and along the north side of the Angara Shield in the Soviet Union. Together these were affected by the great Hercynian orogeny which threw up new fold mountains across ancient Europe, Siberia and China. The Late Palaeozoic (Alleghenian) orogeny of the southern Appalachians was an integral part of these events. As with the Caledonides, fold belts developed along the sites of major sedimentary basins that received thick accumulations of dominantly marine sediments.

Massive Old Red Sandstone strata, Wilderness Quarry, Forest of Dean, England.

The Hercynian Mobile Belt

Reconstruction of the detailed history of the European part of this story – which is to do with events in the 'Hercynian' Mobile Belt – is made particularly difficult by the later 'Alpine' orogeny, which affected most of Southern Europe and superimposed its effects upon those of the earlier events. There is a lot of uncertainty surrounding the subject of lithospheric plate movements during the Late Palaeozoic, but the evidence suggests that ocean separated the European and African continents before the Variscan orogeny. This ocean was slowly closed as two plates, one bearing Europe, the other Africa, collided during the early part of Carboniferous times. During these events the rocks were not only folded, metamorphosed and invaded by granite magmas, but were severely faulted.

The southern margin of the European continent was bordered by a series of major sedimentary basins in the Late Palaeozoic. To the east, similar basins faced the ocean separating Europe from Siberia, the site of the future Ural Mountains, while equivalent basins existed on the eastern side of the Uralide basin bordering the western edge of Siberia. Further basins appear to have existed along the margins of the ocean which separated ancient Siberia from China, while others existed along the west margin of the Pacific and spread down into what is now Indonesia. These basins received thick accumulations of sediments; those along the northern side of the Hercynian Mobile Belt were washed off the eroded Caledonian Mountains of Europe – although when these were eventually lowered so much that they ceased to provide much debris, during the Late Devonian, the main source of debris appears to have shifted to the south where, perhaps, high plateaux existed, suggesting that the Caledonian Mountains had by then been largely worn down.

The Making of the Hercynides

The Devonian and Carboniferous marine deposits are predominantly shales and sandstones with just a few limestones and volcanic rocks. The eroded material appears to have built up in a marine basin which shallowed northward toward the continental platform. Corals flourished in the warm shallow waters of the nearshore. Northward

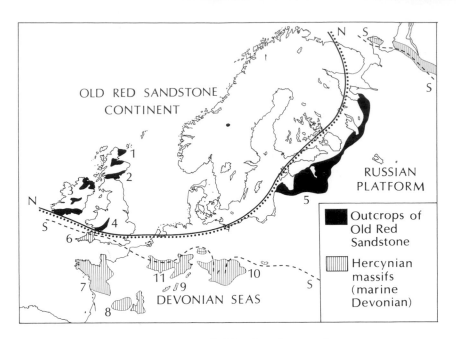

Map showing the outcrops of continental Old Red Sandstone and marine Devonian rocks in western Europe. Old Red Sandstone: (1) Orkney; (2) Moray Firth; (3) Ireland; (4) South Wales; (5) Livonia Hercynian 'massifs' of marine Devonian rocks; (6) southwest England; (7) Brittany; (8) Massif Central; (9) Black Forest and Vosges; (10) Bohemia and Silesia; (11) Ardennes. The line N-N indicates the furthest limit to which the Devonian sea advanced over the Old Red Sandstone continent, while the line S-S shows how far south Old Red Sandstone interfingers with marine strata.

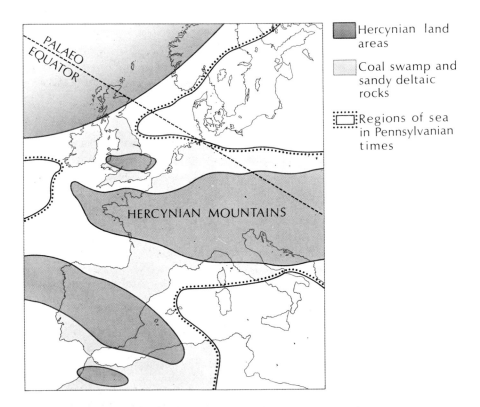

Map to show the Hercynian landmasses in Pennsylvanian times, together with the probable areas of marine deposition. Marginal coal swamp and sandy deltaic deposits are shown in grey. The palaeo-equator ran through the northern tip of Scotland, the Low Countries and Hungary at this time.

again lay the continental interior, sometimes called the 'Old Red Sandstone Continent' on account of the continental strata or 'red beds' (conglomerates, sandstones and shales) that collected on it under the prevailing tropical climatic conditions.

At the end of Devonian times, similar 'red beds' were being deposited in the Ardennes region of France which, at this time, was an island massif rising from the sea; similar massifs extended into other parts of France and Germany. The sea itself seems to have retreated somewhat toward the close of the period, but then advanced in Early Carboniferous (Mississipian) times. It was during this phase that the extensive Carboniferous Limestone deposits of Western Europe accumulated. These are packed with remains of corals, brachiopods, crinoids and other marine organisms. The first two, in particular, serve as zone fossils to work out the detailed geological history.

During the time these rocks were accumulating, Gondwanaland and ancient Europe were slowly converging as the ocean that separated them closed. This implies the existence of a subduction zone, but it is not completely certain whether it dipped northward beneath Europe or southward under Gondwanaland. It has been suggested that the volcanic rocks of the region may just tip the balance in favour of the former arrangement in which an island arc existed above the subduction zone off the southern margin of Eurasia during Devonian times.

The main orogenic phase occurred in the Early Carboniferous, though parts of southern Germany and Bohemia were affected by disturbances in Late Devonian times. During the principal orogeny, the entire mass of sediments and volcanic rocks was intensely folded and metamorphosed. The fold mountains so formed were then intruded by large volumes of granite, which have been dated, and evidently formed in two phases – the first between 350 and 330 million, the second between 290 and 280 million years ago. This orogeny brought an end to marine sedimentation in this region and raised up a new mountain range which extended right across Europe.

The Coal Swamps

During the Late Carboniferous or Pennsylvanian period, erosion of the continent continued, and huge volumes of sand and mud flushed out into the marginal seas. Massive deltas built out and extensive swamps covered these and the wide basins set between the new mountains and the cratonic lowlands further north. In these accumulated the valuable coal seams upon which the strength of the European economy rested for so many years. A similar situation arose also in the Appalachian belt, with which the Hercynian of Europe was then continuous, there being no intervening Atlantic Ocean. The precise method of coal formation will be described in more detail when we look at the Late Palaeozoic story of Gondwanaland.

Events Elsewhere

The main phase of uplift and deformation in the Uralide belt was later, in the Late Carboniferous to Mid-Permian, and faulting in this area continued right through into the Triassic. In the region of the future Atlantic Ocean, rifting allowed large volumes of Permian basalts to rise to the surface and these are now exposed in the Oslo region of Norway and the Midland Valley of Scotland.

During the Permo-Triassic period, most of Europe and northwest Africa was a land area, with non-marine red beds being laid down over wide areas. Climatic conditions were warm and sometimes very dry. This situation found a parallel in North America at the same time. Finally, in northern Europe, two brief incursions by the sea led to the deposition of important salt-bearing deposits (evaporites).

There were many similarities between Europe and North America throughout the Permo-Carboniferous. For instance, the plant remains and reptilian fossils are very alike, and red beds of similar age occur on both continents. Carboniferous coal swamps are common to both. The main episodes of folding occurred at approximately the same times. All this provides impressive evidence for a continuous mobile belt prior to Triassic times. Relatively recent work in West Africa has also thrown light on the situation, for here, too, there is a mobile belt of Permo-Carboniferous age, among the rocks of which are large eastward-thrust slices of deformed strata. The events here could have parallelled what was happening on the northern side of the Appalachian Mobile Belt at this time.

The ancient Siberian shields were also bordered by marine basins. While the Uralide Mobile Zone ran along the western edge of this continent, this was not the only one: thick sequences of marine sediments and volcanic rocks are found in a belt south of the Angara Shield, and running roughly west-east through the southern tip of Lake Baikal. Red beds typify the northern part of this belt, these ranging in age from Devonian to Early Carboniferous; southward they pass into marine limestones and shales. All of these strata were folded toward the close of Early Carboniferous times; further deformation also took place in Permian times. Similar mobile zones also surrounded ancient China; these were aligned roughly along the older Early Palaeozoic basins. One of these – the Central Asia basin – was raised up into a major fold-mountain chain during Late Palaeozoic times, presumably as ancient China and ancient Siberia collided.

Events on the continental interior of Siberia were not dissimilar to those prevailing in Britain. In the west, Devonian red beds accumulated, these passing eastward into marine limestones. The later Carboniferous and Early Permian rocks are mainly limestones, but there was also a significant accumulation of Permian coal. The coals contain a plant flora that is rather like the Glossopteris flora of Gondwanaland, but palaeomagnetic data indicate they were formed at rather lower palaeolatitudes than the latter. Later in Permian times there was widespread volcanic activity over the western part of the continent. The story in China was rather similar.

Mineralization and Plate Tectonics

By the time the Variscan orogeny was ended, most of Europe's mineral wealth had been created. In recent years it has been realized that the creation of this source of wealth has close ties with plate movements and orogenies. Although some metals are concentrated within stationary bodies of magma, simply by the sinking of dense metal-bearing oxide minerals (as is the case with many chromium and platinum deposits), most ore deposits are concentrated rather differently.

Other metals, for instance copper, tin and molybdenum crystallize from solutions associated with granite magmas. Most sulphide ores appear to have been deposited from hydrothermal solutions, which simply are hot fluids that have the capacity to carry metals in solution. Such fluids are believed to originate in magmas and then to rise through the overlying crustal rocks, carrying with them dissolved

Continental Old Red Sandstone rocks from the Midland Valley of Scotland.
TOP Coarse conglomerates containing large cobbles of mainly quartzite.
ABOVE cross-bedded sandstones and pebble beds indicative of deposition in shallow water under the influence of strong-flowing currents. These are typical 'molasse' type sediments. They typify that phase in an orogenic cycle which immediately follows the raising up of a new mountain range, and during which the new mountains are rapidly attacked by erosion.

metals from the magma, or attacking the rocks through which they travel, leaching out certain elements and depositing others. These metal-charged fluids are called mineralizers.

Plate Boundaries and Ore Deposits

Some of the world's richest sulphide deposits are located under the Red Sea. This formed along the side of a former intercontinental rift at which developed the divergent plate boundary along which the African and Arabian plates are spreading apart. The ores are found at a depth of around 2000 m below sea level, in sediments that are anything from 20 to 100 m thick. The concentrated metals include iron, zinc, copper, lead, silver and gold. The pore spaces between the mineral grains of the sediments are saturated with brines that carry the same metals in solution. These metal-charged brines presumably circulated through the sediments above the spreading axis, which is the site of considerable magmatic activity.

The Red Sea is a relatively young ocean; it represents the earlier stages of a continental split. What, we may reasonably ask, happens in the vicinity of a major spreading axis, like a mid-oceanic ridge?

Ridge rocks are altered by heated sea water in nearly every site that has been satisfactorily investigated. It seems that hydrothermal solutions rise from depth, penetrate fissures adjacent to the spreading axis, dissolving metals out of the underlying rocks as they rise upward, and finally concentrating them in ore deposits. Sea-floor sampling shows that iron, manganese, copper, cobalt, chromium, uranium and mercury are enriched in such localities.

One particularly valuable ore concentration is found among the Troodos Mountains of Cyprus where ancient slabs of oceanic lithosphere have been thrust onto the continental margin, then buried by more recent deposits. Study of the ophiolites allows us to investigate, above sea level, a slab of oceanic crust that must have originated at a spreading axis. The valuable copper, iron and chromium ores of the region are closely related to the volcanic rocks in which they occur. Thus, interbedded with the volcanics are manganese- and iron-rich layers that are identical to the mineralized sediments dredged from modern mid-oceanic ridges.

Metalliferous deposits of different kinds are also concentrated along convergent plate boundaries, where oceanic lithosphere is being (or once was) subducted beneath continental crust or an island arc. The processes that operate under these conditions to concentrate ores are complex and very diverse. The valuable ores of the Philippines, Japan and the Cordilleran belt of the western Americas are located on such boundaries. Furthermore the majority of gold deposits, such as those of Alaska, Zimbabwe and west Australia, are found on the sites of ancient convergence zones. The simplest way to explain such occurrences is to assume that mineralizers arise from the melted subducted slab as it plunges down toward the mantle.

Hydrocarbons

The formation of petroleum and other hydrocarbons may also be indirectly linked to the activity of lithospheric plates. Rifting and divergence of continental margins have been shown to be capable of providing all the key factors for the generation and accumulation of petroleum. In such a situation, rifting and subsidence allows a narrow sea to form within a continental block. This sea will have rather

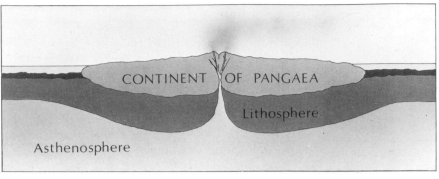

CONTINENT OF PANGAEA

Lithosphere

Asthenosphere

a

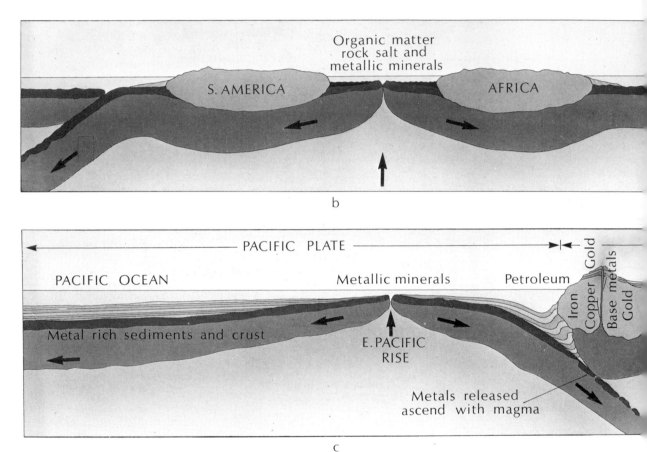

Organic matter
rock salt and
metallic minerals

S. AMERICA

AFRICA

b

PACIFIC PLATE

PACIFIC OCEAN

Metallic minerals

Petroleum

Iron

Copper

Gold

Base metals

Gold

Metal rich sediments and crust

E. PACIFIC
RISE

Metals released
ascend with magma

c

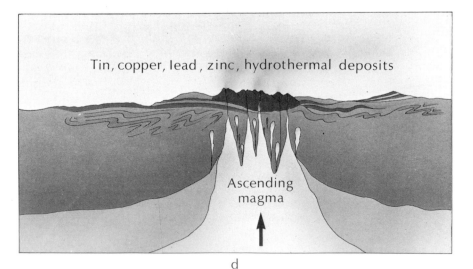

Tin, copper, lead, zinc, hydrothermal deposits

Ascending
magma

d

restricted circulation, providing suitable conditions for the preservation of hydrocarbons within the sediments. Furthermore, if the basin is in a subtropical or tropical site, the rate of evaporation may be sufficient for rock salt to be formed – as happened in the Hercynian belt and in the early stages of the Red Sea. As the new sea widens, so the evaporites and organic matter become buried by younger sediments. The high heat flow associated with the rise of magmas may help the organic material to yield petroleum and the salts will form domes or 'diapirs' which act as very efficient traps for hydrocarbons.

The rich oilfields found around the coasts of Britain are an example of hydrocarbon generation in just this kind of circumstance. The North Sea formed part of a major Jurassic rift system predating the spreading of the North Atlantic ocean. Sediments accumulated in the rifts, spreading beyond them as the basin continued to subside. The shallow, restricted sea which covered the area deposited organic-rich muds which provide the source for much of the oil. Later tectonic movements associated with further development and eventual spreading of the North Atlantic rift system formed a variety of trapping structures. As the hydrocarbons were expelled from their source rocks under a combination of heat and pressure, they rose upwards until trapped. Some of the traps in the southern North Sea are formed by the rise of salt diapirs from the Late Permian evaporite deposits of the Zechstein Sea.

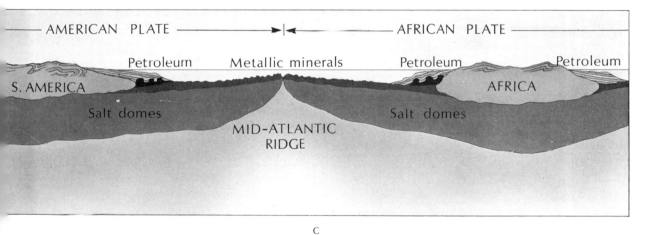

c

The relationship between plate boundaries and mineralization. In (a) the ancient continent of Pangaea pulls apart, thick layers of salt and organic matter accumulating on the receding plate margins. Uprising currents beneath the spreading centre cause sea water to become an ore-forming mineralizer, leaching metals out of the hot crustal rocks deep down and eventually precipitating them on the sea floor and in sediments flooring the oceanic crust. In (b), as the sea floor spreads salt domes that originated in the thick evaporite layers rise up through the sediments of the continental margin. In so doing they trap oil and gas which are generated from organic matter preserved within the sediments. In (c) the Pacific plate is being subducted beneath the western margin of the American continent. Sediment and crustal material enriched in metallic ores by activity along the East Pacific Rise are partially melted. Metals rise with the magma that is generated, to produce the metal-bearing rocks of the Andean chains. In (d), hot spots rising within lithospheric plates are the source of metals such as tin, lead, copper and zinc.

17

GONDWANALAND EVOLVES

Gondwanaland in the Early Palaeozoic

Palaeomagnetic data reveal that during the Early Palaeozoic, present-day southern continents (South America, Africa, Australia, India and Antarctica) formed the continent of Gondwanaland. This occupied a rather different position with respect to the poles than did the other continents and the most reasonable conclusion to be drawn from this is that it was separated from them at the time. This next chapter traces the evolution of this supercontinent, not only in Palaeozoic but also in more recent times. The break-up of Gondwanaland left widespread evidence among the rocks of various modern continents and had a profound influence on Mesozoic and Cenozoic events. The painstaking piecing together of the various shreds has enabled geologists to write the story that follows.

The General Picture

During the earlier part of the Palaeozoic, the supercontinent appears to have been virtually surrounded by subsiding basins. In these were laid

Gondwanaland in the Early Palaeozoic, showing mobile belts in grey and cratons in yellow.

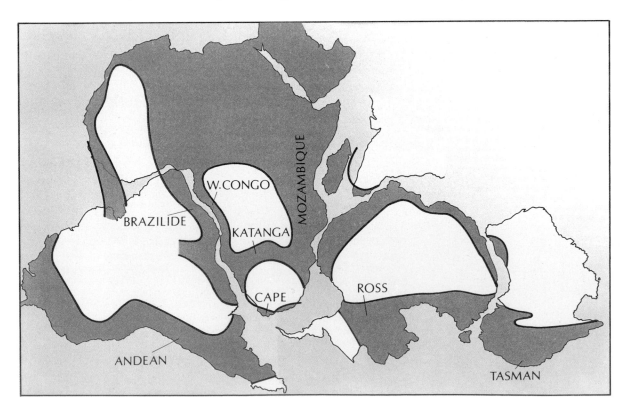

down sedimentary rocks which are now found in such widely separated places as eastern Australia, New Zealand, northern India, North and West Africa, Florida, the Andes, South Africa and Antarctica. Large remnants of the ancient cratonic core which formed the foundation of Gondwanaland, are located in central Australia, northern Pakistan, India, Brazil, Africa, Central America and parts of the southern United States. Palaeomagnetic and palaeoclimatic evidence from the rocks of the continental interior shows Gondwanaland stretched across a wide range of latitudes during Early Palaeozoic time.

The Continental Interior

Among the sediments that accumulated on the continental interior are distinctive sandy horizons which contain both faceted and striated pebbles and boulders. These tillites are of Ordovician age and have been found not only in the Saharan region of northwest Africa, but also various parts of South America. The probability is that these are glacial sediments and that the boulders were dropped from huge continental ice sheets which spread across the African and South American Shields. This interpretation is supported by geophysical data which show that during the Ordovician the South Pole lay in west central Africa. All of the Ordovician glacial rocks were apparently located within 50 degrees of the Late Ordovician South Pole.

The central part of Australia was a major depositional basin and there were smaller ones on the northern and western edges of the continent. The rock succession found in these includes sandstones, shales, limestones and salt deposits or evaporites. Similar evaporites are found within basin sequences in northern Pakistan, where in the Salt Mountains are 400 m of salt and gypsum deposits. Evidently both India and Australia experienced an arid climate during the Lower Palaeozoic, a conclusion confirmed by palaeomagnetic information and contrasting sharply with glaciation in Brazil and Africa.

The southeastern part of the United States and southern Mexico were both part of Gondwanaland during the Early Palaeozoic. Wells drilled in southeast Alabama, southern Georgia and northern Florida have revealed the existence of Lower Palaeozoic volcanic and sedimentary rocks here also in which the fossils are similar to those found in North Africa.

Margins of the Supercontinent

At the beginning of Palaeozoic time, Gondwanaland was girdled by subsiding basins that evolved into active organic zones. The shallower nearshore waters were the collecting grounds for clastic sediments and shelly limestones and the living environment for abundant shelly organisms which thrived there. The deeper offshore regions were characterized by black shales and turbidites, together with volcanic rocks derived from offshore island arcs, while the fauna was dominated by the planktonic graptolites. The remnants of this system are now located in widely separated continental locations and provide us with vital evidence in favour of continental drift.

Part of this circum-Gondwana orogenic system now occupies the Andean chain of South America. In Early Palaeozoic times the more westerly of these South American basins collected thick marine sediments and volcanic rocks but volcanic rocks are not found further east. In places the succession is 3000 m thick. Glacial rocks of Early Ordovician to Mid-Silurian age are reported from western Bolivia and southwest Argentina. Intense orogeny deformed the basin sediments toward the end of the Early Palaeozoic, and some geologists believe orogeny may have been promoted by the subduction of oceanic lithosphere beneath an island arc or arcs that sat off the continental margin, with eventual collapse of those arcs.

Another part of the mobile zone can be traced through Antarctica, along the axis of the Transantarctic Mountains. Cambrian clastic rocks, limestones and volcanic rocks are here found to sit more or less conformably on a Late Precambrian erosion surface. Orogeny deformed these rocks somewhat earlier than those of South America, the granites associated with the orogenic disturbances being dated at between 450 and 520 million years. Eastern Australia also has widespread Cambrian deposits. Thick carbonate sequences are found in the west of this zone and among these are algal stromatolites which are believed to have built reefs in the shallow waters near to the cratonic margin. Similar rocks are found also in New Zealand. Sedimentation appears to have stopped abruptly during the Middle Cambrian, and the strata were folded and metamorphosed.

Another portion of the circum-Gondwanaland belt runs through the Himalayas, Pakistan, Iran, Turkey, Greece, Yugoslavia, Italy and North Africa. Between Cambrian and Devonian times this series of troughs was joined with those that bordered the western side of Africa. In this zone few volcanic rocks are found but clastic sediments may attain a thickness of 5000 m in places. This tract was not affected by deformation during the Palaeozoic, and evidently plate convergence did not affect this border of the supercontinent.

Lower Palaeozoic marine sedimentary rocks are found also along the coastal regions of West Africa and in the southern Appalachians. In the latter region, 10,000 m of shales and volcanic rocks are found in the belt extending from Carolina through Virginia to Georgia. The same rocks doubtless occur elsewhere but are covered by younger strata. The volcanic rocks in these troughs may have been produced within island arcs, and it is possible that a subduction zone plunged southeastward beneath the supercontinent in this area.

The Late Palaeozoic Ice Age – 1

The Late Palaeozoic was a time of profound change. Not only were there distinctive climatic trends, but also there was a converging together of the continents until, about 280 million years ago, one great supercontinent – Pangaea – existed. Before that time Gondwanaland had been the largest of three major continents and had extended from the Equator to the South Pole. Laurasia, situated in northern latitudes, was then in two parts, these being apparently separated from each other by a wide sea that included several large islands. The westerly one was Laurentia. The more easterly of these masses is sometimes called Angaraland. An eastward-broadening ocean separated the latter from Gondwanaland. This has become known as the Tethys Ocean.

We saw in an earlier section how on the North American craton there were repeated fluctuations in the relative level of the sea and the land during Late Carboniferous (Pennsylvanian) times. Similar evidence comes from Gondwanaland at this time. During the Pennsylvanian there were repeated incursions and withdrawals of the sea, reflected in the cyclic deposits found on the borders of the ancient oceans surrounding the supercontinent. Very reasonably, it has been postulated that these rhythmic fluctuations occurred in response to

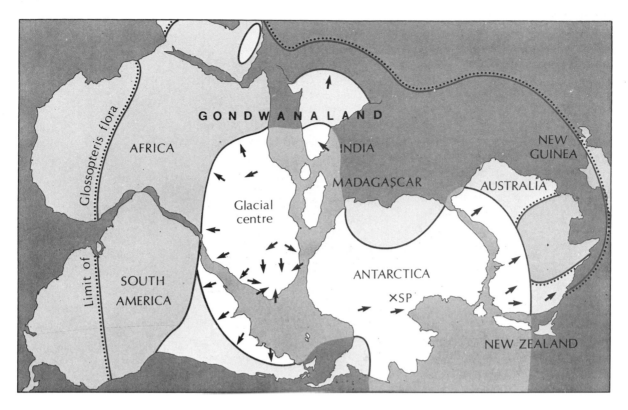

GONDWANALAND

AFRICA

Glossopteris flora

Limit of

SOUTH
AMERICA

Glacial
centre

INDIA

MADAGASCAR

ANTARCTICA

×SP

NEW
GUINEA

AUSTRALIA

NEW ZEALAND

The Late Palaeozoic ice sheets of Gondwanaland. The centre of glaciation shifted during Permo-Carboniferous times; thus South America and southern Africa were affected first and eastern Australia and Antarctica last.

changes induced by pulsatory ice-sheet growth, centred on the Late Palaeozoic South Pole, which then lay within Gondwanaland. We know that very rapid sea level changes have occurred during the last (present?) ice age, therefore there is every reason to suspect that such would have happened in the past.

What was the world like during this period? It was certainly quite different from today. For a start, the land surface was substantially less luxuriantly clothed in vegetation during earliest Carboniferous times, although there was a rich flora beneath the waters of the oceans. It seems likely that climatic zones were established and they were not vastly different from today's, although overall temperatures may have been slightly lower. Then, in Pennsylvanian times, as the various cratons moved toward one another, the situation changed. By the middle of Permian times, the sea separating the two components of Laurasia had been eliminated, the new supercontinent gradually rotating until it collided with Gondwanaland, fusing them together into one vast mass: Pangaea.

The collisions generated new mountains along the sites of the old sedimentary troughs. The creation of these lofty barriers, together with the broader geographical changes, appears to have had a profound effect upon the ocean currents, atmospheric circulation and the world climate. Extensive volcanism accompanied orogenic activity, so that great sheets of lava were spread over the margins of the lithospheric plates. There is evidence that snowfields capped high mountains, not only in mid-latitudes but also in the equatorial zone. All of these factors conspired to bring about a global cooling and the growth of extensive ice sheets over much of Gondwanaland. Glacial deposits and the fossilized imprint of glacier activity can be found in such diverse places as South America, Antarctica, southern Africa, India, Arabia and Australasia. These occurrences provide us with one of the most powerful pieces of evidence for the existence of Gondwanaland.

The Late Palaeozoic Ice Age – 2

There seems little doubt that there were alpine glaciers in both Bolivia and Argentina as early as Devonian times. Subsequently they became much more extensive and there were at least two major glacial episodes during the Carboniferous period. In Brazil roughly 1500 m of glacial beds accumulated during the period 320-270 million years ago, detailed study of the glacial beds showing that there were at least

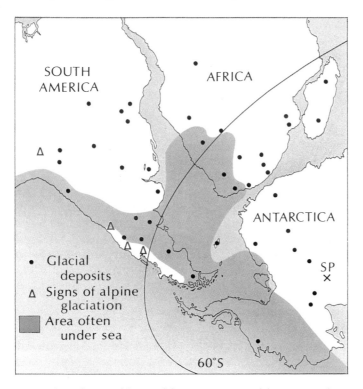

Map to show the possible site of the maritime part of the great ice sheet. Over most of this region the base of the ice sheet was close to or beneath sea level, rendering it rather unstable.

seventeen glacial-interglacial cycles during the 50 million year time span. When features produced by the moving ice are studied – for instance, glaciated rock pavements and striated valley walls – they indicate that the main ice sheet lay off the eastern coast of South America and the direction of ice motion was from east to west. This may seem odd at first sight, but when we take into consideration the configuration of the continents, it becomes more realistic.

The African craton was adjacent to South America at this time and in southwestern Africa there are significant glacial deposits of Permian-Carboniferous age. The most famous of these is the Dwyka Tillite, which has been traced across southern Africa into Madagascar. The glacial rocks are over 1000 m thick in the Transvaal, which appears to have been one of the main areas of ice dispersal. In the higher part of the succession are plant-rich beds that contain a flora dominated by the seed fern, Glossopteris, but which also contain the remains of Mesosaurus, a Permian aquatic reptile found only here and in Brazil.

Glacial rocks of the same age are found also in Australia where the earliest effects of the changing climate were felt in the mountains of Tasmania and southeastern Australia. Subsequently the ice spread as far as northern Queensland. The widespread occurrence of tillites suggests that most of southern Australia must have been covered by ice at least once during this period. Glacial features left behind after these glaciers retreated strongly suggest that the main ice movement was from the south. At the time, Antarctica occupied this position. For a while this posed a problem as no glacial deposits of comparable age were known from that continent. In 1960, however, extensive glacial strata were located in the Transantarctic Mountains and elsewhere. In places these reach a thickness of 900 m. Such a massive occurrence emphasizes how Antarctica was a major ice source during Permian-Carboniferous times.

India and Pakistan must be included in this story, too. The northern margin of India was a part of the south coastline of the Tethys Ocean during the Ice Age. Upland regions further south were apparently the source for extensive ice sheets, which sometimes spread into the sea. The glaciation here spanned the interval 310 to 270 million years ago and there were at least three principal glacial episodes. Glacial deposits of similar age are found also in the Falkland Islands.

Of necessity the details of the glaciations have very much been simplified. It is, however, widely accepted that southern Africa was glaciated for the longest period; that as Gondwanaland slowly shifted with respect to the poles, so the focus of glaciation changed; and that the most reasonable interpretation of the palaeomagnetic and palae-ontological evidence is to assume that South America and south-western Africa became glaciated very early on. The focus then shifted to South Africa, Malagasy (Madagascar) and India, finally moving on to Antarctica and Australia.

The Rise of Plants

Putting all the evidence together, it seems that in the Late Palaeozoic Tropics it was warm enough for the development and survival of tropical swamp vegetation near sea level. Away from the oceans,

Reconstruction of a Carboniferous forest. This would have been dominated by seedless trees and seed ferns.

ABOVE LEFT *Glossopteris* (Gondwana realm; Spain).

ABOVE RIGHT *Lonchopteris rugosa* (Amerosinian realm; Spain).

however, arid deserts were the rule. It is possible that glaciers pushed to within 30 degrees of the Equator in Early Permian times, so there is likely to have been a compression of the climatic zones. Life endeavouring to survive in regions marginal to the great ice sheets would have had a particularly tough time and only the more resilient species would have survived the rigours of glaciation.

Enormous reserves of coal have been located and exploited in North America, Europe and Asia. These were laid down in widespread swamps during Pennsylvanian and Early Permian times. Similar coals exist on the southern continents, at this time a part of Gondwanaland. While the former apparently enjoyed warm climatic conditions, the latter developed in a much cooler environment. The principal ingredient in all of these deposits were the enormous numbers of land plants which grew and diversified at this time. Study of the fossil record shows how primitive vascular plants had evolved by the Late Silurian period (about 400 million years ago). By Late Devonian times these evolved to form extensive lowland forests of tree ferns with a canopy about 10 m high. They appear to have spread quite rapidly over the northern hemisphere. By the beginning of the Carboniferous period they were the chief contributors to peat and swamp vegetation. Fossilized remains suggest these attained a height of about 20 m in some cases.

The earlier, more primitive plants reproduced via spores, but the seed ferns which flourished during the Carboniferous were the first to reproduce by seeds. These were to become the stock from which the later cycads, conifers and flowering plants developed during the Mesozoic era. The most important contributors to the coal deposits were the Lycopsids and Sphenopsids which stood tall in the coal swamps of the times. The former is represented today by the low-growing temperate-forest club moss, while the latter's only descendant appears to be the curious little horsetail, Equisetum.

The various continental floras were quite diverse, and sufficient is known to recognize different plant 'realms'. Thus the floras found in the region of the Angara Shield of ancient Siberia are preserved in coals that were deposited at around latitude 45 degrees north. Palaeomagnetic data shows that although these are similar in some respects to plants recovered from the coals of Europe and North America, the latter were deposited within 10 degrees of the Equator. The Glossopteris flora of Gondwanaland, on the other hand, lived in latitudes higher than 30 degrees south. The three realms are, therefore, the Amerosinian Realm (which includes the floras of the equatorial belt), the northern Angaran Realm, and the southern Gondwana

ABOVE LEFT *Nothorhacopteris argentinica* (Gondwana realm; Argentina).

ABOVE RIGHT *Angaridium finale* and *Gondwanidium sibiricum* (Angara realm; USSR).

Realm. There are significant differences between these three realms, the restriction of the Glossopteris flora to Gondwanaland being another of the vital pieces of evidence for continental drift and the subsequent break-up of the Gondwanan supercontinent.

The Coal Measures

Pennsylvanian times were characterized by the deposition of coals on all of the continents. They were laid down on the margins of Gondwanaland and they were also deposited in inland basins among the Hercynian massifs of Eurasia as well as on the continental margins. The Appalachian region also received its share of coals. As we have suggested, the type of cyclic deposition during which most Carboniferous coals were laid down, may well have been closely related to up-and-down movements of sea level in response to variations in the volume of ice locked up in ice sheets during the Late Palaeozoic ice age we have just described.

To take the British Coal Measures as an example: toward the close of the Carboniferous period, most of Britain and western Eurasia had been levelled by erosion. Huge deltas formed on the margins of the continental masses and there tended to be great uniformity of geographical conditions over huge distances. For instance, one particular marine bed can be traced from Ireland to the Soviet Union! Under such conditions the typical western European Coal Measures cyclic sediments were deposited.

The typical rhythmic sequence, called a cyclothem, begins with a coal seam that formed from the dead growth of dense swampy forest. This is usually overlain by shale which contains marine fossils whose presence indicates that the sea had inundated the low-lying swampy deltas. Upward the marine shale passes into shale with a fauna which clearly lived either in brackish or fresh water; the inference here is that lagoons must have developed. Further up the sequence the shales become more sandy, passing up into a thick sandstone unit usually containing plant remains. The change to coarser material is either taken to indicate that the higher land behind the deltas had become slightly more elevated, leading to enhanced erosion and the production of coarser sediment, or that river-borne sediment was being dumped over the site of the former lagoons due to the increased width of the tidal flats. Higher in the cyclothem, the sandstone usually becomes rather more silty and finally passes up into another coal seam. Immediately beneath the coal there is often what is called a seat earth, which represents a fossilized soil horizon in which the roots of the swamp plants grew. The gradual reversion to coal formation implies

141

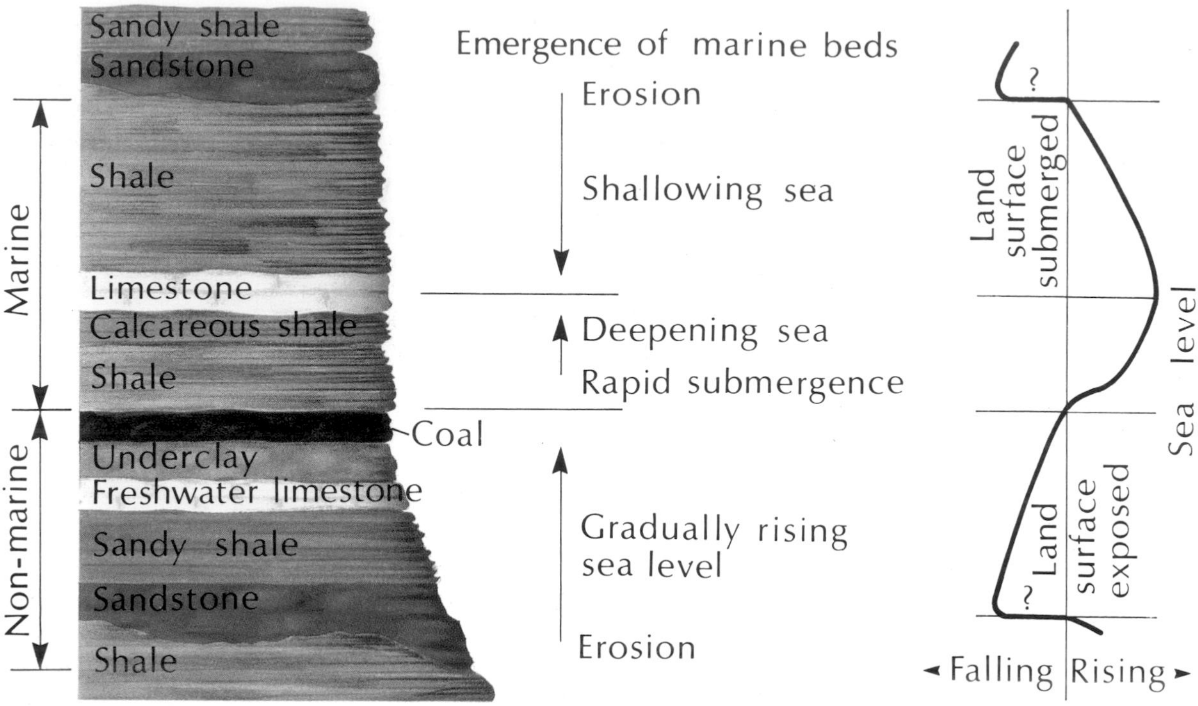
Sandy shale
Sandstone
Shale
Limestone
Calcareous shale
Shale
Coal
Underclay
Freshwater limestone
Sandy shale
Sandstone
Shale
Marine
Non-marine
Emergence of marine beds
Erosion
Shallowing sea
Deepening sea
Rapid submergence
Gradually rising sea level
Erosion
Land surface submerged
Land surface exposed
Sea level
Falling
Rising

Section through a typical cyclothemic unit from the Coal Measures. On the right is a graph relating the different beds to changing sea levels.

that the impetus of uplift gradually died away again, the swamps once more encroaching over the delta flats.

Rhythmic sequences of this type are repeated many, many times within the Late Carboniferous succession. Coal actually accounts for a relatively small part of each cyclothem and if erosion for some reason stripped away the plant debris before it could be buried by later sediment, the coal horizon may actually be absent.

Late Palaeozoic Animals

There were considerable differences between Early and Late Palaeozoic animal life. For instance, after the Devonian period the trilobites became very rare and were extinct by Late Permian times. Snails do not appear to have been very abundant but the coiled ammonites started to increase in numbers and were very widespread in their distribution. They provide us with important zone fossils in Devonian and later Palaeozoic strata. The brachiopods continued to evolve and, in particular, spiny-shelled forms developed. The development of spines assisted in anchoring them to the sea floor and also helped in straining food from the muddy sea water, as well as affording them protection from their enemies.

During Devonian times, reef-forming groups like stromatoporoids, corals and bryozoans were important. However, this changed in the Carboniferous, when crinoids (sea lilies) took over the lion's share of reef-building activities. They became not only abundant but very diversified and have never been more so, even though they survive in reduced diversity to this day. Of the smaller organisms, the protozoans underwent tremendous evolutionary changes. The foraminifera, in particular, diversified and proliferated. Many of the genera evolved very rapidly and were apparently highly specialized, so that they provide palaeontologists with ideal zone fossils.

One of the most significant features of Late Palaeozoic life was the

ABOVE Freshwater lamelli-branchs of Coal Measures age.

CENTRE LEFT Bryozoans, brachiopods and their spines on a slab of Upper Permian shale.

CENTRE RIGHT Pentacrinites – a Jurassic crinoid.

BOTTOM LEFT Gastropods and lamellibranchs in Jurassic ironstone.

BOTTOM RIGHT Devonian fish, Dapedius

colonization of the land. In Late Devonian times the first amphibians appeared. These were fish-like but very quickly developed so that the body became flattened and the eyes moved towards the top of the head, as if they lived largely in shallow water. Some adapted to the land by redesigning their backbones and legs; but they tended to be clumsy and, in any case, had to return to the water to lay their eggs. In contrast, the reptiles, whose eggs had a tough outer skin, were able to survive more readily away from the water. The earliest reptiles looked little different from the amphibians, but during the Carboniferous these quickly evolved and some very bizarre forms inhabited the low-lying swampy plains. One group, the Pelycosaurs, had long spines with skin stretched between them. These were probably used to regulate their body-heat budget in some way. They became so specialized that they could apparently inhabit only a narrow ecological niche, and were extinct by the end of the Permian period. Other groups were much more successful and some of the mammal-like genera gave rise to the true mammals that flourished during the Mesozoic era.

In contrast, there was a rapid increase in the occurrence of insects during Late Carboniferous times. Abundant fossils of large cockroach-like insects and enormous dragonflies have been recovered from lake and lagoon deposits within coal-bearing sequences. Hitherto insects had existed, having first made their appearance in Silurian times, but they appear to have evolved very rapidly during the later period.

18

STIRRINGS IN
THE ALPS

Introduction

The Mesozoic era saw the break-up of Gondwanaland and the development of the Alpine orogenic cycle. This began in Triassic times with the growth of major troughs on or closely alongside the older Hercynian belts and ended, in the European region, with the uprise of the Alps and Carpathians during Tertiary times. Orogenic activity also affected Asia, the Earth's highest mountain chain, the Himalayas, being thrown up somewhat later in time, but during the same general phase of upheaval. Activity in the latter region has continued more or less until the present. The Alpine Cycle therefore takes us from the events of the Late Palaeozoic up to the present day. These orogenic processes were part of the extremely complex sequence of events that accompanied the approach of Eurasia and Africa, with subduction of the intervening oceanic crust.

The European Alps
(Otztaler Alpen, Austria).

The Alpine Cycle

The Alpine Cycle really began during the Triassic, about 240 million years ago, with the deposition of marine shales and carbonate sediments in basins that developed among the eroded stumps of the ancient Hercynian mountains. Triassic and Permian sedimentary rocks are to be found in a broad tract that includes the Alps and Carpathians, one of the most closely studied mountain belts in the world. It was among these mountains that European geologists first began to perceive the way in which a major mountain belt grew and developed.

There is not only evidence that sediments were laid down in Europe during the Triassic Period, but also that carbonate reefs grew in both Austria and northeastern Italy. This contrasts sharply with the continental deposits which accumulated to the north in Germany, Poland and elsewhere, and indicates that the present Alpine area was near the southern margin of the European continent. It is likely that these reefs grew near to an ancient island arc that built up across that region. During the Jurassic period the sea spread northward, flooding much of Europe. It reached its maximum extent during the Late Cretaceous, when it covered the whole of Europe and Russia and extended also over North America, North Africa and Arabia. Without doubt, this was one of the greatest marine episodes in the stratigraphic record. Because it occurred during a subdivision of the Cretaceous called the Cenomanian age, it is widely known as the Cenomanian Transgression.

In the Alpine region during Mid-Cretaceous times the crust became increasingly unstable and a number of elongated basins developed which were separated by large islands. Sporadic crustal uplifts alternated with periods of rapid sedimentation as short pulses of tectonic activity affected different parts of the region. This stage has since been recognized to have occurred during other orogenic cycles, and has become known as flysch sedimentation. Eventually, during the early part of the Cenozoic era, the geography of these islands became more and more different and the shallow seas began shrinking as Europe and Africa approached each other until, during Oligocene and Miocene times, the mobile belt shuddered under the culminating Alpine Orogeny as the continents collided. This threw the buried sediments

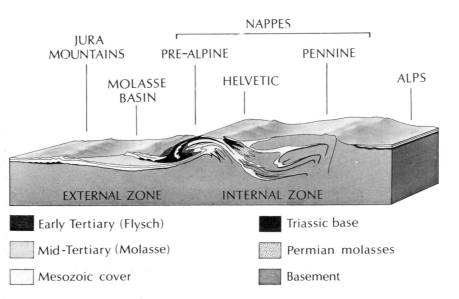

Section through the western part of the Alpine Chain.

into great contortions and raised up a new fold-mountain chain whose remains we can still see today.

The orogeny itself strongly deformed the rocks, not only the Mesozoic sediments but also the Hercynian basement beneath. Magmas rose into the lower part of the crust, giving rise to large bodies of both granitic and basaltic rocks. The crustal uplift that affected what is now Switzerland caused immense slabs of only partially coherent rock to slide over one another and pile up. Where deformation affected more consolidated sediments, the folding was of a more plastic kind, the strata being rucked up into enormous folds which became flat-lying, giving rise to recumbent folds. These flat-lying structures, first described from the Alps, became known as nappes – in many cases the individual fold limbs were torn from their roots and transported tens of kilometres from their original positions in response to the continental collision occurring.

During the main orogenic episode there was regional elevation of the European crust and the sea was expelled from the central Alpine tract. This meant that the products of the ensuing rapid erosion were deposited in non-marine basins adjacent to the rising mountain chain. This rather distinctive phase has been found to be typical of the later stages of an orogeny and has been given the name of molasse sedimentation.

During orogeny the rocks of the more central regions were strongly metamorphosed. Radiometric dating shows that the majority of the Alpine nappes had been formed by about 70 million years ago and the metamorphism shortly afterwards. These orogenic processes were a function of the elimination of the oceanic crust and bringing together of the rocks trapped on the margins of the Eurasian and African cratons, as they moved inexorably toward one another on their respective lithospheric plates.

Since the orogeny finished, the Alpine chains have been deeply eroded; by Pliocene times much of the region was probably of quite low relief. The magnificent mountain scenery of today is largely the result of uplift and erosion just before the Pleistocene.

Generalized map of the Alpine Chain and its eastward extension into the Caucasus and Zagros Moutains of Iran. The complex structure of the Mediterranean region indicates that substantial rotations of parts of southern Europe must have occurred during Tertiary times and since the main fold belt was produced (1 = Adriatic Massif; 2 = Rhodope Massif).

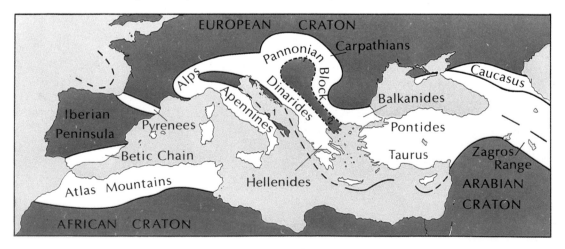

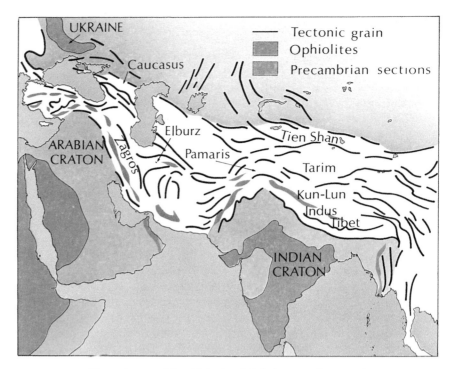

ABOVE Tectonic map of the 'Alpine' mobile belts extending into the Middle East and Central Asia.

Flysch Deposition

During the early stages of the cycle, sedimentation along the continental margins tended to be dominated by chemically precipitated rocks, like limestones. Later, however, as the conditions became less and less stable, the nature of the sediments changed dramatically, and land-derived debris became the dominant sediment, this either becoming mixed with the carbonates, or entirely replacing them. This change occurred in Cretaceous times, and the resultant deposits are of Late Cretaceous and Palaeocene age. Their formation was the signal that orogeny had begun in the core of the mobile belt. This type of deposit, known as flysch (from a German dialect word), is seldom thicker than 3 km in the Alpine region, but it is very widespread. In different places, it is of different age (so is termed a diachronous deposit). Thus the earliest flysch was generated where the earliest orogenic disturbances were felt. Since the disturbances began at different times in the various parts of the Alps, the flysch deposits vary in age. The younger flysch sequences are located in the exterior zones, that is, those regions furthest from the orogenic core, where they may be of Palaeocene or even Eocene age.

The main source of flysch was the rising orogenic belt itself. Thus as folds developed and the crust thickened, the more elevated regions were rapidly attacked by erosion and the detritus strewn out quickly by streams and rivers along the sea shore. Very rapid deposition meant that a complete jumble of different rock types, coarse- and fine-grained, was deposited side by side. The whole would have been very unstable. Earthquakes generated by continuing orogenic activity set in motion

LEFT Current markings or 'flute casts' on the undersides of turbidite bedding plane, North Wales. Each lobate cast represents the infill of a current scour mark. The shallower end of individual casts indicates the 'downstream' direction.

147

the loosely consolidated layers which slid down submarine slopes into deeper waters, there to spread out as turbidites.

Certain features found within the rocks can be used to trace in which directions the predominant currents flowed. These 'markers' suggest a very complicated pattern of deposition, as if short-lived ridges within or on the margins of the various basins, contributed not only sediment but also to a complex regime of currents. In both the Carpathians and the Hellenic ranges, redeposition of the flysch debris was largely accomplished by the activity of turbidity currents – the turbid submarine flows that produce typical turbidite sequences – which flowed axially along the major basins.

Molasse

The upheavals of Cretaceous times largely terminated the production of flysch deposits, since they had the effect of eliminating the surviving areas of ocean, except in parts of the Aegean. By Oligocene or Early Miocene times, large upland tracts had been elevated, much of this being accomplished by block faulting. These youthful mountains were soon attacked by erosion, and the debris produced was transported either into low-lying basins on the margins of the old mobile belt, or among the new mountain massifs. Some of these old basins, now of course filled with sediment, underlie the European lowlands and marine gulfs of today. The sediments that did accumulate in this post-orogenic phase are called molasse deposits.

The molasse is usually unconformable upon folded, metamorphosed rocks which once were deep down in the fold-mountain belt. It thus clearly postdates the climactic orogenic phase. Much of the sedimentation was like the Coal Measures in the sense that it was cyclic, and cyclothems are a feature of some areas. Typical molasse includes non-marine conglomerates, sandstones and shales, while coal seams are developed in the cyclic sequences. Brackish-water or marine sediments sometimes interfinger with these rocks indicating that they were deposited in marginal basins. In some zones, however, marine rocks dominate: thus the Middle Miocene rocks found along the northern edge of the Alpine-Carpathian belt are typically marine in the area east of Munich, and they form the greater part of the molasse sequences in the Eastern Mediterranean and Adriatic. In complete contrast, elsewhere – as in the Balkans, Cyprus and Yugoslavia – evaporites accumulated in Miocene times.

Molasse, then, is typical of the later stages of an orogenic cycle. We can thus liken the Alpine molasse to the Old Red Sandstone, which was laid down in the post-orogenic phase of the Caledonian cycle we outlined earlier. Similar deposits have been found to occur among the successions of most orogenic cycles and, like flysch deposits, are useful marker horizons.

The European Craton

During Late Palaeozoic and Early Mesozoic times the climate appears to have been hot and arid over much of this region. Strata of Permian-Triassic age are widespread and these include fossil sand dunes, as well as other typical continental deposits laid down in rivers and lakes that periodically dried up. The reddish coloration such rocks acquire has led to their being called New Red Sandstones. Not all of the region was desert, however, and there is clear evidcence that shallow seas covered parts of the continent from time to time.

Molasse-type pebble beds and sandstones.

The marine incursions inundated parts of both central and north Europe. Late Permian strata, including a rather thick series of dark bituminous shales called the Kupferschiefer, evidently accumulated under a major body of water, and this has become known as the Zechstein Sea. The Kupferschiefer or 'copper shales' are of particular importance, since they became impregnated with copper and provided Germany with an important source of copper ore. Another important economic commodity won from these rocks is salt, for important evaporites were laid down and the shallow sea repeatedly dried up under the scorching Sun. Indeed, the saline concentration became so high that the teeming life which had characterized its waters was wiped out as thick beds of rock salt, anhydrite, potash and magnesium salts were laid down. The famous Stassfurt Mines have been carved through these rocks. Much of this salt eventually rose as diapirs (domes) into the denser Mesozoic and Tertiary rocks covering the craton and these are also found beneath the Low Countries and southern North Sea, where they form important traps for hydrocarbons. The Zechstein Sea did not connect with the Tethys Ocean, but was actually an arm of the Arctic Ocean.

Somewhat later, during Mid-Triassic times, the Tethyan Ocean, which lay to the south of the craton, expanded and shallow water covered much of Germany and the Low Countries, parts of France, Spain and North Africa. Even at this time, however, the massifs of the Ardennes, the Vosges and the Massif Central stood proud. The Baltic Shield, Bohemia and Britain also were not covered by its waters. The typical calcareous deposits of this sea are extremely fossiliferous, and are known as the Muschelkalk.

At the close of Triassic times, the cratonic mass of Europe lay to the north of the relatively deep Tethys Ocean which lay between North Africa, southern Europe and southern Russia. The continent was the

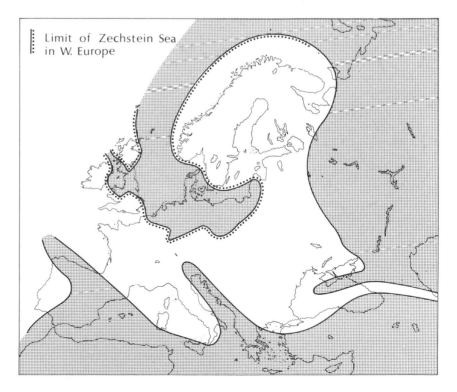

Limit of Zechstein Sea in W. Europe

Map to show the Carboniferous and Permian seas that invaded the European craton. The main marine area covered much of Russia and extended through Turkey into Greece, Spain and North Africa. An arm did, however, extend into north Germany and Britain. This has become known as the Zechstein Sea and in it, distinctive dolomitic limestones of Permian age were deposited.

ABOVE Early Jurassic sandy beds (Bridport Sands), Dorset, England. Similar sandy beds occur at different levels in the Lower Jurassic succession in southern Britain. These shallow-water deposits are typical of the shoreline deposits of the sea that spread widely over western Europe.

TOP RIGHT Liassic limestones and mudstones, Warwickshire, England. More typical of the shallow 'epicontinental' Jurassic sea are these rapidly alternating grey mudstones and limestones. Rich in marine fossils, particularly ammonites, they attract fossil collectors in large numbers. Zoning of this part of the Jurassic sequence is accomplished largely by the use of the cephalopods.

CENTRE RIGHT White chalk cliffs, Denmark. The chalk, a pure organic limestone, was laid down over vast areas of Europe and America during the great Cenomanian Transgression. This inundated both the European and North American cratons and was undoubtedly the most widespread event of its kind in the geological record.

levelled remnant of the ancient Hercynian Mountains which had been transformed into a vast desert. During the Jurassic period, the seas invaded the southern border of this great landmass a number of times. Because it was of such low relief, little coarse detritus was available, and the characteristic sediments were calcareous muds, called marls, and organic limestones. These collected in large sedimentary basins, one of which is known as the Anglo-Parisian Basin, currently bisected by the English Channel. Others developed in the French region of

BELOW The Muschelkalk Sea which spread over the European deserts during Triassic times. Its animal population has a number of special characteristics (it was rich in individuals but poor in species) and seems to have derived from the deeper Alpine Sea to the south. Land areas shown in brown, marine deposits in green.

Aquitaine, in northeastern Spain, over southern France and also over the regions which now are the Black Forest and Bohemia.

Later, in Cretaceous times, there was a more widespread invasion of the sea and many of these Jurassic basins accumulated marine Cretaceous deposits. Indeed, during the early part of Late Cretaceous time, in the Cenomanian Age, the sea flooded almost the whole of Europe and Russia south of the Baltic Shield and much of northern Africa. It also invaded North America, lapping up against the Appalachians and spreading over the continental interior. It was the first and last time during the Phanerozoic that the Sahara itself was inundated – indeed it was, without doubt, one of the greatest marine transgressions of all time. Beneath the waters prevailed a remarkable uniformity of conditions, under which was laid down the famous 'Chalk', the soft white limestone of the White Cliffs of Dover. This is composed of millions of tiny skeletons of algae that flourished in the warm shelf seas.

At the beginning of Tertiary times, central Europe was a low-lying land area. The Baltic Shield, to which the Caledonides of Britain and Scandinavia were strongly welded, behaved as a positive region, but the old Hercynian tracts were much fractured, buried by sediments and only parts of them stood out above the general low level. About 60 million years ago, the sea slowly inundated a vast area as it spread from the west until, by Late Oligocene times (about 25 million years ago), it had reached southern Russia. Later, during the Miocene, it slowly withdrew again and by the Late Pliocene had apparently departed from the continental interior altogether.

The Crust in Upheaval

The cause of the Alpine Orogeny, as we have seen, was the head-on clash between the Eurasian and African plates. This gradually eliminated the intervening ocean and brought together thick sedimentary belts that had lain both to the north and south of the line of contact, and also involved the crystalline basement beneath. Although this explains the broad features of the Alpine chain, it does not readily answer all of the geological problems posed. It has been found necessary to consider the possibility that as well as the two major lithospheric plates involved in the collisions, there were also a number of smaller microplates that got mixed up in movements adjacent to the principal collision zone, and moved independently of Eurasia and Africa.

The Mediterranean poses a considerable problem for interpreters of geological history, as here the arrangement of the various mobile zones is very complex and geophysical data suggest the story is unusual. Palaeomagnetic results from rocks ranging in age from Palaeozoic to Triassic show that those of Corsica, Sardinia, north Italy and Spain have undergone large rotational movements with respect to the main European craton. For instance, the Italian peninsula seems to have been rotated about 43 degrees clockwise in the Late Palaeocene, then 25 degrees in the opposite direction since the Mid-Eocene!

Complex movements thus appear to have played an important part in the history of the Mediterranean region. High among the fold mountains of southern Spain are detached folds (nappes) which must have been transported from a source region now to the south where there is now deep water. Amid the Apennine Mountains of Italy are clastic sedimentary rocks that derived from land somewhere southwest of the Italian peninsula. This is rather difficult to believe! When, however, we realize that Sardinia and Corsica once occupied this position, and that the Tyrrhenian Sea – which currently separates Italy from both Corsica and Sardinia – opened due to the clockwise rotation of Italy, the situation clarifies.

Although of the same age, the Pyrenees do not form a part of the Alpine fold belt. In fact, they appear to have originated in a tremendous rucking-up of strata covering the cratonic basement, followed by mass sliding of the rocks over one another, as movements occurred in the Hercynian basement beneath. The more central regions of this range seem to have been elevated at least several kilometres, while the basins to the north and south were filled with both flysch and molasse of Tertiary age.

Then, there are the Apennines, a still active part of the great Alpine mobile belt. As we have seen, these may well have been rotated from a position much further west, so that what is now the western side of the tectonic unit may once have faced toward the north. Recent work suggests that west of Rome, in Elba and Corsica, are ophiolites which represent fragments of oceanic crust that bordered the colliding plates during earlier stages of the Alpine orogeny.

As the Alps rose, in response to the convergence of Europe and Africa, so the Mediterranean – once opened to both west and east – began to close. Closure first took place in the east during the Miocene, as Europe and Arabia collided. Then, a little later (perhaps in the Late Miocene) palaeontological evidence suggests that the western end also became closed off. The result of this sealing-off of the sea was widespread deposition of evaporites during the Messinian age (about 6 million years ago).

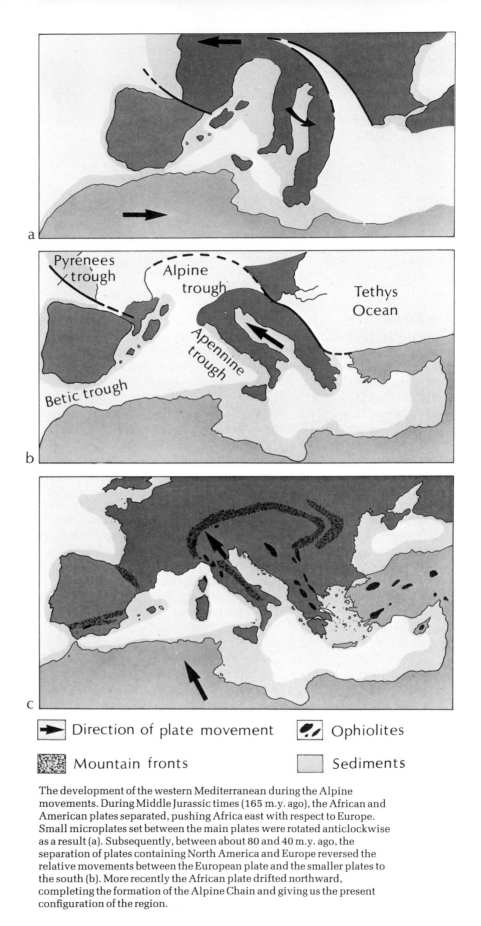

Direction of plate movement

Mountain fronts

Ophiolites

Sediments

The development of the western Mediterranean during the Alpine movements. During Middle Jurassic times (165 m.y. ago), the African and American plates separated, pushing Africa east with respect to Europe. Small microplates set between the main plates were rotated anticlockwise as a result (a). Subsequently, between about 80 and 40 m.y. ago, the separation of plates containing North America and Europe reversed the relative movements between the European plate and the smaller plates to the south (b). More recently the African plate drifted northward, completing the formation of the Alpine Chain and giving us the present configuration of the region.

Karakoram ranges, Tibet.

The Himalayas

The Alpine mobile belt passes through the Middle East and then swings northeastward and merges with Earth's greatest mountain range: the Himalayas. The Tertiary sedimentary rocks of Turkey and Iran comprise a thick succession of clastics which were eroded from uplands situated on the northern side of the Tethyan Ocean. To the south they pass into carbonate rocks, deposited in shallower waters. The main folding of these rocks was accomplished during Miocene times, when the Zagros Mountains of Iran were uplifted as a result of collision between Eurasia and Arabia, then a part of the African plate.

The Himalayas themselves are just part of a great complex of mountains and plateaux that occupy vast areas in Central Asia, Mongolia and Tibet. There is at least 2,500,000 km^2 of terrain over 4 km above sea level in this massif. The scattered geophysical data that exists for this part of the world suggests that the continental crust is roughly twice its normal thickness here, i.e. 70 km instead of the usual 35 km.

Only the south and west portions of this belt are directly related to the Alpine orogeny that shook Europe. The ranges to the north and east are built from faulted remnants of the rocks of Precambrian and Palaeozoic orogenic belts which were affected by upward block movements in Late Tertiary times. These movements were specifically linked with the collision between the Indian subcontinent and Eurasia, which began about 40 million years ago. After the initial contact, India decelerated but continued to drive northward into Eurasia for another 2000 km. It is not surprising the effects are so marked!

The colliding plates in this episode were both continental in character and, since this type of material does not subduct readily, the pressures of collision had to be absorbed in some other way. Part of the stress was relieved by the stacking up of huge slices of rock which were scraped from the driving front of the Indian craton. These slices, which were moved roughly west-east along low-angled thrust faults, included rocks from both mobile zones and the basement beneath. The sideways movement of these rocks was at least in part due to the fact that the converging continental margins were non-parallel, thus driving the first promontories to have come in contact in this direction. It is from these rocks that the Himalayas were formed. A further part of the compression appears to have been taken up in the upfaulting of large tracts, to give major topographic features such as the Tibetan Plateau. Even so, together these movements could only account for about a half

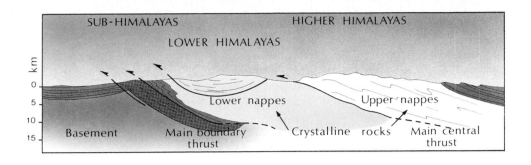

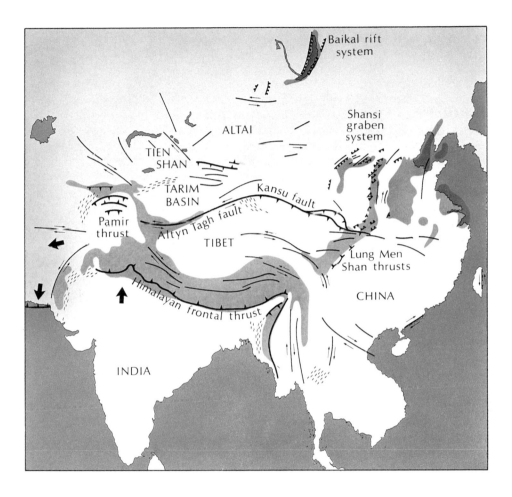

▨ Extension zones	⊥⊥ Thrust fault
▨ Uplift regions	⌒⌒ Normal fault
⌗ Folds	⇌ Strike-slip fault

Africa and Asia collide to produce the Himalayas and the great Tibetan Plateau. Between 60 and 40 m.y. ago plates containing Eurasia and Peninsular India collided. India then slowed down but nevertheless continued to thrust northward for at least 2000 km. The fact that buoyant continental crust does not subduct meant that crustal pressures were absorbed, not so much by subduction, but by the stacking up of vast overthrust slices of rock, which once were northern India, on to one another. More of the horizontal compression was taken up by the uprise of the vast Tibetan Plateau, while China was pushed forcibly eastward along enormously long strike-up faults.

of the required 'give'. Thus it has been suggested that much of the convergence was absorbed by China, which was pushed eastward, out of the way of India, along enormously long strike-slip faults.

It is possible to discern, with a little ingenuity, the line of demarcation between rocks originally belonging to the southern and northern plates. This line is known as the Indus Suture as it follows the line of the River Indus in the northeast of Pakistan, and across toward the River Brahmaputra in Bangladesh. The folded rocks produced during the climactic phase of orogeny lie to the north and west of this and form the great ranges of the Pamirs, Karakoram and Hindu Kush. To the south lie stacked huge slices of crust from the Indian craton, from beneath which undeformed India emerges south of the mountains.

The geological cycle whose record we can uncover in the northern ranges of the Himalayas took considerably longer to run its course than did the corresponding Alpine events of the southern margin of Eurasia. In fact, there are no major tectonic breaks to be found between the Proterozoic and the Late Mesozoic strata which were deposited just before the start of the orogeny itself. The Himalayan flysch deposits – characteristic, remember, of active orogeny – span the interval between the Mid-Cretaceous and Eocene, indicating that orogeny at the southern margin of the Eurasian continent began in Late Mesozoic times, perhaps 95 million years ago. The non-marine molasse deposits were laid down significantly later than in the Alps – in Late Miocene, Pliocene and Pleistocene times. As in the Alps they partly postdate the major orogenic episode, yet even these young deposits have been deformed and thrust beneath the Southern Himalayas.

Events in Siberia

The Urals were periodically rejuvenated during Mesozoic and Tertiary times and, as a consequence, these ranges expose old Precambrian and Palaeozoic rocks. The Urals were marginal to the craton which had become stabilized in Precambrian times and has, since that time, received a variety of shallow-water or continental deposits which, in the Verkhoyansk and Khakingat basins, attain a thickness of over 6 km and span an interval of 1500 million years.

During Late Permian and Early Triassic times there were extensive outpourings of basaltic lavas over the northwestern part of the Siberian craton. This great igneous event had no parallel on the European craton and it rivalled in extent the ancient Keweenawan lavas of North America. The lavas and their associated feeder dikes cover an area of 1,500,000 km² and are in places 1 km thick.

Basalt is not of course a characteristic rock of continental cratons and was extruded as a result of stretching of the continental crust, which allowed magma to rise from the upper mantle regions. Crustal fracturing and disruption is also suggested by the occurrence of diamond-bearing volcanic pipes – typically produced during episodes of explosive activity. These are found near to the basalt region, but somewhat further east and are mainly Triassic in age, like the basalts.

During Jurassic times the sea invaded much of southern and eastern Siberia, but the great Cenomanian Transgression, which deposited the Chalk in Europe, did not cover a significant area of Siberia. Instead there was the deposition of immense deposits of coal, presumably under conditions resembling those of the European and North American Carboniferous swamps though in temperate rather than tropical latitudes.

Lake Baikal, Siberia. The lake occupies a rift in Cenozoic and younger rocks.

Part Four

GONDWANALAND AND MORE RECENT EVENTS

The coastline of northern California – meeting place of lithospheric plates.

19

GONDWANALAND BEFORE THE BREAK-UP

The Unbroken Continent

By the close of Palaeozoic times, the continental interior of Gondwanaland had become stable and experienced little orogenic disturbance. The Equator crossed through the northern end of the Red Sea, across the Sahara, thence through southern Florida. On the west and south, a series of marginal basins ran along the western side of what is now South America, through western Antarctica and into eastern Australia. Along the northern perimeter, the southern margin of the Tethys Ocean ran through North Africa, Arabia and northern India.

We have already sketched out the Late Palaeozoic history of Gondwanaland, describing how, during Pennsylvanian and Permian times, there was widespread glaciation of the interior. Many of the glacial beds are overlain by Permian continental deposits, and incursions of the sea are indicated by marine members within some successions. The accumulation of continental deposits continued in some parts of the continent right through into Early Cretaceous times: a period of about 150 million years, showing how stable conditions were over very large areas. Laurasia and Gondwanaland were almost certainly joined in Triassic times, the northern coast of South America being in contact with eastern Central America, northwestern Africa with the eastern margin of North America, and southern Spain with North Africa, and it is probable that a vast land area existed.

The Rocks of Gondwana

Late Palaeozoic-Mesozoic sediments accumulated thickly in certain basins and were widespread over the Gondwanan cratons. Most of the deposits were non-marine, but occasional marine incursions deposited a few marine bands and these have proved extremely useful for correlating the strata from one basin to another. The sequences found on the various cratons of all of the continents that then were a part of Gondwanaland show remarkable similarities, as we might expect if they were all close to one another.

The cratonic sequences – which are all extremely similar – span the period Late Carboniferous-Cretaceous. The lowest strata of the Gondwana successions – the Carboniferous beds we have already discussed – include those glacial deposits that indicate that much of Gondwanaland was situated at high latitudes and experienced a glaciation no less severe than the Pleistocene Ice Age. The overlying Permian sequence includes abundant coal measures that contain the unique Glossopteris flora. Higher in the succession these beds give way to yellow, green and brown sandstones and marls containing abundant reptilian remains. Study of these rocks and their fossil fauna and flora indicates that Gondwanaland experienced a gradual warming of its climate, this

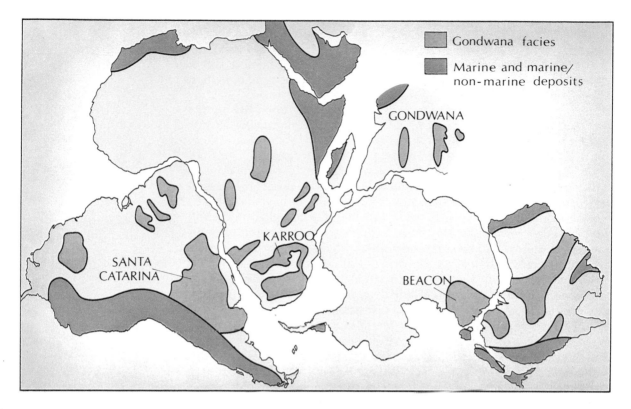

GONDWANA

KARROO

SANTA CATARINA

BEACON

The rocks of Gondwana. This map shows the distribution of Permian and Early Mesozoic non-marine formations (called the Gondwana facies) on the various continents which once formed Gondwanaland. The accumulation of sediments of continental facies started during the Late Palaeozoic glaciation and continued in many places until the Jurassic or even Cretaceous period. Thus in some basins a record of continuous non-marine deposition spanning 150 m.y. is found.

reaching a peak in Late Triassic times, about 215 million years ago. Those parts of the supercontinent now forming Africa and South America appear to have drifted slowly away from the polar regions toward the tropics. The Triassic sandstones, so typical of all basins, were parallelled by similar aeolian rocks (the New Red Sandstones) in Laurasia, which was then positioned within the northern tropical belt.

Shales, fine-grained sandstones, alluvial deposits and some limestones accumulated in the cratonic basins. Such rocks typically collect in relatively shallow water. The rate of sedimentation must therefore have been finely balanced against the rate at which the basins themselves subsided, so that the depth of water did not change a great deal over lengthy periods. Of the one or two marine incursions that did occur, one particular marine band, of Early Permian age, was rich in the remains of the marine shell Eurydesma, an indicator of cool marine conditions. This horizon, known as the Eurydesma band, is found in both southwestern Africa and Brazil, and has proved to be a useful means of correlating the successions in the two continents.

Mesozoic Life

We have signalled already that the later part of the Mesozoic was a time of biological change. Many of the invertebrate animal groups that flourished during the Mesozoic had disappeared altogether by the close of the era. Perhaps the best example is that of the ammonites, which became extremely abundant during the Jurassic and Cretaceous, then slowly dwindled in abundance. Plants, too, saw significant changes, the most marked of which was the evolution of flowering plants (angiosperms). These first made their appearance in the Early Cretaceous. By Late Cretaceous times, these had taken over from their earlier cousins, the gymnosperms, of which the modern conifers are representatives. Fish also underwent changes. The earlier Mesozoic forms still had heavy scales and skeletons built mainly from cartilage; but by the close of the era, small scales and bony skeletons were the rule, as they are in modern fishes. The amphibians continued to decline as time went by, and the reptiles saw dramatic changes, the best-documented being those relating to the dinosaurs.

During the Early Triassic there was a wealth of reptilian life on the continents and from one group, the thecodonts, were descended the famous dinosaurs, and also (later) the birds and crocodiles. These thecodonts and their relatives ruled the land for over 150 million years, starting out as small lizard-like animals, but from them eventually developed the well-known huge heavy-tailed beasts that could walk on two legs. The more primitive forms seem to have been cold-blooded like living reptiles; however, detailed physiological studies of their remains have suggested to palaeontologists that some of their later descendants had warm blood, and were presumably more active than once thought.

More advanced dinosaurs can be divided into two principal groups: those whose bone structure points to a reptilian mode of travel (saurischians or 'lizard-hipped'), and those that were 'bird-hipped' (ornithischians). Both appear to have been warm-blooded and there is some evidence to suggest that many were capable of quite rapid movement. One of their more pronounced evolutionary trends was toward gigantism: some really were monsters, being at least 15 m in length and standing to comparable heights. Not all, of course, were this large; some were more dog-like in proportions. All had extremely small brains for their size.

The saurischians, in particular, evolved the very large forms we have all seen depicted in popular books. Brontosaurus and Diplodocus are but two of these monsters. Some of them must have been among the fiercest carnivores ever to be seen on Earth. Ornithischians, on the other hand, were mainly herbivorous, but these also grew to very large size. The armoured Stegosaurus is probably the best known of this group; this flourished during Jurassic times. Another member, Triceratops, had strong horns with which it presumably could attack its enemies with some vigour. This form and its close relatives roamed about in large numbers during the Late Cretaceous and was among the last of the dinosaurs. By the close of the Cretaceous they were extinct.

Dinosaurs also lived in marine environments while some learned to fly. These latter were the precursors of our modern birds, although the exact evolutionary link is rather obscure. Mammals also evolved during the Mesozoic, probably developing from reptiles during the Triassic. They may well have had a diet of insects and smaller reptiles.

We have already mentioned Mesozoic plants. Palm-like cycads achieved prominence during the Jurassic and Early Cretaceous, while the seed ferns declined. Conifers and gingkos proliferated, and the flowering plants made their first appearance. The earliest of these appears to have occurred during Early Cretaceous times. These flowering plants underwent their greatest proliferation during the Jurassic and Early Cretaceous, just as the ruling reptiles were giving way to the mammals. The causes of these dramatic changes are still being debated. We shall return to this topic later.

The Beginning of the End

Outpourings of plateau basalts, mainly during Jurassic times, signalled the beginning of the end, not only for Pangaea but also for Gondwanaland, which was to fragment completely before the close of

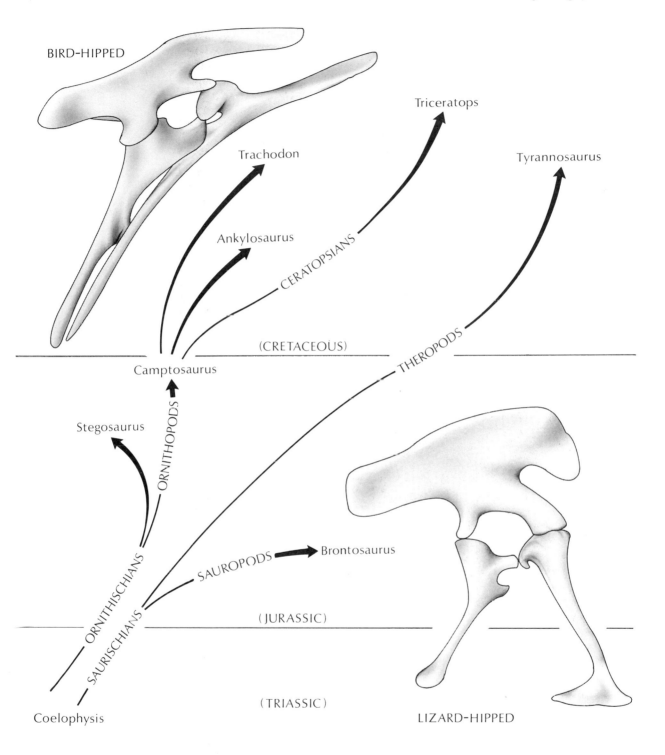

BIRD-HIPPED

Triceratops

Tyrannosaurus

Trachodon

Ankylosaurus

CERATOPSIANS

(CRETACEOUS)

THEROPODS

Camptosaurus

ORNITHOPODS

Stegosaurus

SAUROPODS → Brontosaurus

ORNITHISCHIANS

SAURISCHIANS

(JURASSIC)

(TRIASSIC)

Coelophysis

LIZARD-HIPPED

Diagram showing two orders of dinosaurs. These developed from a common ancestor but had distinctly different pelvic structures. The Ornithischian branch ('bird-hipped') differed from the Saurischian ('lizard-hipped') in the arrangement of the hip structure as shown at the bottom of the diagram.

Eocene times. We must now state what is the evidence for believing that in Triassic times, a plate carrying Gondwanaland began to separate from the plate bearing Laurasia. At the same time as Late Triassic sediments were being deposited in fault-bounded basins that had developed along the east coast of North America and northwest Africa, basaltic magmas were rising toward the surface in both regions. Radiometric dating of these lavas and dikes from both continents show that they crystallized between 192 and 202 million years ago, which corresponds to the period Late Triassic to Early Jurassic. Along the eastern side of North America rather more alkaline magmas rose into the crust at roughly the same time. Thus there was much igneous activity during the Late Mesozoic.

If we look at the rocks of the ocean floors we find that the oldest known sediments to be laid down on the oceanic crust that developed between North America and Africa were recovered from near the Bahamas Bank by the oceanographic research vessel *Glomar Challenger*. These proved to range in age between 162 and 151 million years (Late Jurassic). Similar sedimentary rocks were found some 530 km from the margin of the North American continent east of North Carolina; these gave a radiometric age of 155 million years. Knowing the distance between two magnetic anomalies (one dated at 153 million years and another at 107 million years) recorded in the sea floor of this region, geophysicists estimated that between 153 and 107 million years ago the sea floor spread at an average rate of about 1.1 cm per year. Using the formula *distance = spreading rate × time*, it can be seen that the time that must have elapsed since the spreading began equals 530 divided by 11 (11 being the number of km travelled by the sea floor in every million years, i.e. 1,100,000 cm per 1,000,000 years). This gives an answer of 48 million years, which puts the time of the initial separation at about 203 million years ago, in the Early Jurassic.

Palaeomagnetic data show that the oldest magnetic anomaly on the floor of the central Atlantic, then separating North America from Africa, has an age of 153 million years. This anomaly is located 535 km from the eastern margin of the North American continent so, again assuming a spreading rate of 1.1 cm per year, the two continents would have started to separate about 202 million years ago.

The *Glomar Challenger* also collected deep ocean sediments from near the centre of the Gulf of Mexico which, at the time of separation, would have been on the line along which the separation of North America and South America took place. These proved to be of Late Jurassic age and showed that separation between these two continents must have begun well before this time.

Studies of the way in which the continents have moved with respect to the magnetic poles also add weight to the view that break-up started in the Early Jurassic. Thus curves plotted for Africa and North America agree well during Carboniferous-Triassic times, but for Jurassic and later time they begin to diverge more and more, showing that the continents went their separate ways from this time. The relationship between North Africa and Europe is less clear. They must have separated, and some evidence suggests they began to move apart about 180 million years ago. Other geologists would put the time of separation somewhat earlier. We favour the Late Triassic as the most likely time for the initial separation of Gondwanaland from North America: this final severing of ties between Laurasia and Gondwanaland began the slow fragmentation of Pangaea, supercontinent extraordinary.

The Plateau Lavas

Just before the break-up there was a significant episode during which enormous volumes of basaltic lavas were poured out. These emerged from fissures in the cratonic crust, which opened in response to crustal stretching. Sills and dikes of dolerite – the coarser-grained intrusive variety of basalt – filled countless millions of such fissures, which fed and extended well beyond the limits of the flows themselves. This magmatic activity was very clearly a function of what was happening to the crust as the supercontinent began to fragment.

Because they spread over such large regions, giving rise to flat-lying topography that has been etched out into countless plateau-like landforms, these lava fields are known as plateau basalts. In northern India, plateau lavas form the Deccan Traps, a sequence of lavas that reach a thickness of 3000 m and cover an area of over 550,000 km^2. They were poured out during Late Cretaceous and Early Cenozoic times. Similar rocks are found in Brazil where there are at least 6000 m of volcanic and non-marine sedimentary rocks of Late Cretaceous age. Basalts occur also in southern Africa, where the Upper Gondwana Basalts cover large areas of the eastern region; furthermore there are countless related sills and dikes over a much wider area of the craton. Interbedded with the flows are sediments containing Jurassic plant remains. Radiometric dating confirms this date for the volcanism.

Also featuring in the African story are 'kimberlites' – rocks that occur in narrow pipes, and appear to have formed in response to explosive volcanism from very deep levels. The kimberlites themselves contain minerals like olivine, pyroxene, mica and garnet and are rich in the chemical element magnesium. They include large numbers of fragments of other rock types, most of which appear to have been formed at very high pressures, presumably deep within the crust or upper mantle layer. It is within these pipes that diamonds are found.

Some of the African kimberlites are Jurassic, although they mostly seem to be of Early Cretaceous age. In Antarctica, too, basalts are very widespread and yield a radiometric age of 160 million years, putting them firmly in the Jurassic period. Basaltic sills and dikes are common also in Tasmania, but the earliest of these are slightly older, dating from the Late Triassic but extending into Jurassic times.

The principal basaltic activity seems, therefore, to have been concentrated in Jurassic-Cretaceous times, but the basalts are not the only products of this active period. Other igneous rocks, rich in the alkali elements potassium and sodium, have been located in all of the southern continents, save for Antarctica. These rocks are usually coarsergrained than typical basalt and occur in ring-shaped masses, many of which appear to be part of larger volcanic centres. Such rocks are typical of areas of continental crust that are thinner than normal, generally due to rifting. They are of comparable age to the basalts.

Igneous activity of Jurassic-Cretaceous age was, therefore, a feature of all of the Gondwana cratons. Basaltic volcanism does not, however, normally affect the interiors of continents and so this episode must point to unusual conditions. The fissure-related volcanism developed as the continental lithosphere experienced stretching, giving rise to extensive rifts up which magmas rose, to flood out onto the cratons. Thus, although this volcanism took place within the Pangaean continent, it marked the sites of oceans which were soon to develop and separate its now widely scattered remnants.

20

GONDWANALAND DISRUPTED

The Quiet Period Ends

The great outpouring of plateau lavas that occurred between 160 and 120 million years ago signalled the end of Gondwanaland as a supercontinent. So much basaltic lava was exuded that those parts of the crust that experienced stretching were additionally depressed by the lava burden. In these thinned regions of continental crust, basins formed and these were invaded by the sea. Much of this evaporated in the hot climatic conditions, so that salt deposits formed. These evaporites can be found deeply buried along those margins of the southern continents which represent intra-Gondwanan rifts. Gradually these marginal basins subsided even further and the salt water of the Tethys Ocean gradually invaded them. Initially these new seas were both shallow and rather narrow, rather like the modern Red Sea. Nevertheless, they were the predecessors of the modern Atlantic and Indian oceans, even if their beginnings were humble.

The southward expansion of the Tethyan Ocean can first be traced as it passed between eastern Africa and India, about 160 million years ago. Gradually the waters linked with those already in existence in the rifts around both Madagascar and South Africa. Ten million years later, the ocean similarly spread up the eastern coast of peninsular India. The ocean began to separate South America and Africa at some time between the Late Jurassic and Early Cretaceous, 150-130 million years ago. More recently, about 80 million years ago, the ocean came between New Zealand and Australia, which then was still attached to the western part of Antarctica. The rifting apart of the latter took place 40 million years later. The expansion of the oceans and oceanic crust and the movement of the continents have continued until the present day.

The East African Rift Valley

The great East African Rift Valley crosses the main crustal doming of the craton, which lies well to the east of the continental axis. It is the major part of a 5000 km long tract of fracturing that extends from the Limpopo in the south to the Red Sea and Jordan Valley in the north. The southern end of the rift appears to have begun forming during the Mid-Tertiary times. The main phase of faulting, however, is of Miocene-Pliocene age. The accompanying igneous activity that began in the Miocene has continued to the present.

The rift is a region of high heat flow, and seismic and volcanic activity are characteristic. Vertical displacements along the main faults are considerable; thus in the regions of Lake Nyasa and Tanzania, the floor lies between 2 and 3 km below the rim. Spectacular fault scarps are seen at the edges, the main faults being normal faults produced by tension within the continental crust. While there are modest accumulations of Mesozoic and younger sediments on the rift floor, it is the volcanic rocks that dominate the stratigraphy. Igneous activity started during the Mesozoic in the south and then spread northward. The main rift splits into western and eastern forks, the former running through Uganda, the latter through Tanzania and Kenya. A host of explosive volcanic centres pierce the floor of the western branch, the magmas usually being rich in the alkalies, particularly potash. Peculiar lavas, called carbonatites (which are predominantly carbonate rocks) are typical in this region. The very explosive character of some of these volcanoes shows that volcanic gases played a very important role as the lavas rose to the surface.

The floor of the eastern branch, together with the borders of the rift, are flooded by plateau lavas. These are particularly evident in Ethiopia and northern Kenya and include basalts, phonolites and trachytes. The huge central volcanoes of Mount Kenya, Meru and Kilimanjaro lie on the shoulder of the rift further south. The chemistry of their lavas is similar, but among their rocks are more alkaline types, too. These massive volcanoes contribute to the spectacular scenery.

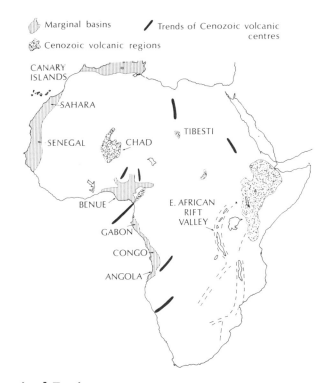

RIGHT The marginal basins of the African continent. Many of these formed in response to subsidence of the edges of the craton along ancient fractures. During Cretaceous and Tertiary times, substantial accumulations of sedimentary rocks collected in them. Similar basins developed at the edges of the other continents.

BELOW Flood basalts of Columbia River Plain.

The Marginal Basins

During Jurassic, Cretaceous and Tertiary times, marine sediments collected in large basins along the perimeters of each of the ancient cratons that had formed Gondwanaland. These sediment-filled basins are known technically as 'aulacogens'. Toward the continental interiors, the sediments of the basins interfinger with shelf or continental rocks. The rocks that did collect in these marginal basins have proved to be immensely important since they are often valuable oil reservoirs.

In many cases the new basins run along the sites of much more ancient mobile zones, some of which formed as early as Late Precambrian times. This relationship is very clearly seen along the Atlantic

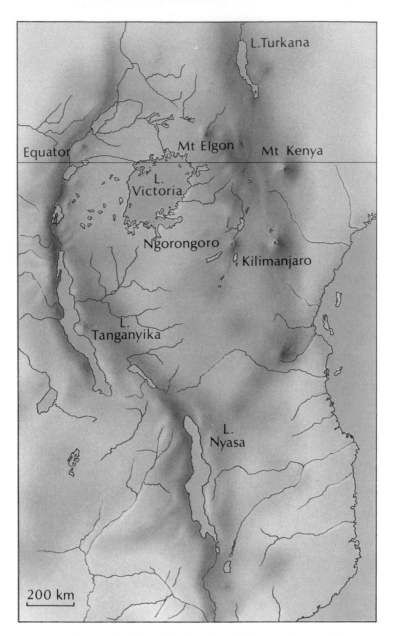

The East African Rift Valley. This huge downfaulted region extends through the eastern side of the African craton from the valley of the Limpopo River, through Ethiopia and eventually joins the Red Sea-Jordan Valley fracture belt. In all this great fracture system has a length of 5000 km.

coast of Africa where a number of large basins are to be found. One of these, the Senegal Basin, overlies the ancient Mauritanide mobile belt which had been active until the Late Palaeozoic. During Cretaceous and Tertiary times, marine sediments collected here, the fill becoming 5 km thick! Many of the rocks are limestones, and the Cretaceous rocks are pierced by large salt domes. This indicates that evaporites were among the earliest of the basin sediments: a feature common to many of these new basins.

Another basin, the Benue, extends inland from the Niger Delta. This has a width of at least 200 km. Filling it is a succession of shallow-water sediments of Cretaceous and Tertiary age, 6 km in thickness.

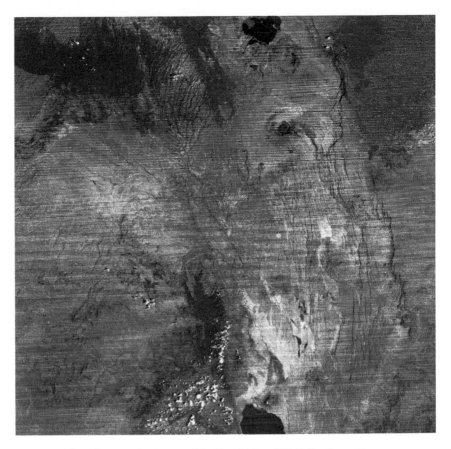

Landsat photograph of a part of the East African Rift Valley in southern Kenya. Lake Natron, situated on the border between Kenya and Tanzania, can be seen at the bottom of the picture, just east of centre. In this false-colour image, vegetation is picked out by reds, and is concentrated on the higher ground. Nairobi itself is situated near the eastern edge of the frame just above centre. Dominating the image is the rift faulting which here runs almost due north-south. A number of volcanoes can be seen on the Rift floor, particularly obvious being Suswa volcano, situated west of Nairobi and south of Lake Naivasha (top).

Among the ancient basement rocks to the northwest of the basin is a series of alkaline granites and volcanic rocks which are enriched in tin minerals. Similar enrichment is found in a line of northeast-southwest trending igneous intrusions that pierce the crust on the opposite side of the basin. Significantly, the latter igneous rocks extend seaward into the chain of volcanoes that ends in the still active São Tomé centre. This illustrates a very close connection between volcanism and these sites of early continental rifting.

Large basins also occur in Zaire and Angola, along the east coast and in Arabia. Along the northern coast of Africa, Jurassic strata, including many limestones, were laid down in a shallow sea that must have connected with the Tethys Ocean to the north. Peripheral basins of this kind are found also on the other continents, and are of comparable age and size. Africa simply provides us with a particularly good example of Late Mesozoic and Tertiary developments.

After the Break – the Pattern of Sedimentation

Marginal basins were a feature of the continental peripheries, but geological processes also affected the continental interiors. We shall commence with a look at Africa, whose Tertiary history is in many ways parallelled by the other Gondwanan continents. The later history of this great continent is dominated by the evolution of the Earth's largest cratonic landmass, as well as the growth of the Rift System.

The north-south dome on which the rifting developed lies well to the east of the axis of Africa. This arch of ancient rocks forms a major watershed, to the east of which the land falls away towards the coast. To the west, however, things are different. Several major 'sags' in the craton received non-marine sediments right through from Palaeozoic times. These interior basins, of which the Kalahari and Congo are the best known, were the collecting grounds for a variety of lake sediments, wind-derived sandstones and a kind of deeply eroded calcareous dust, called loess.

At the margins of the continent, terrestrial deposits interfinger with shallow-water marine rocks, such as sandstones and limestones. Such Late Cretaceous-Early Eocene strata are found along the east coast of South Africa, for instance, and are also seen in Libya, Egypt and Arabia. To the northwest, in the Atlas Mountains and to the northeast, in the Zagros and Elburz ranges, calcareous sediments may be accompanied by evaporites that pass upward into wedges of clastic rocks worn from the newly elevated lands to the south. The great oilfields of the Middle East are found mainly within the Tertiary limestones on the southern flanks of the Zagros Mountains.

More recently, Africa's geological development has been confined largely to periods of uplift and erosion, whereupon huge tracts of horizontal or, at most, gently dipping strata have been hewn into major 'erosion surfaces' that have been dissected by rivers.

South America A similar kind of story can be followed in South America. Here the majestic Andes were eventually to rise along the sites of marginal basins that continued to rise during Tertiary times. These most ancient rocks – those of the Guyanan and Brazilian shields – are situated in the broadest part of the craton and are separated by the huge basin of the Amazon river which has collected not only recent alluvium, but also sediments going back to Palaeozoic times.

The continued rise of the marginal Andean basins along the western perimeter of South America forced the seas to be even more restricted during the Tertiary than they were during Cretaceous time. Thus, during Palaeocene times, marine deposits were confined to a narrow coastal belt running along the western border of the continent. At this time, South America was moving away from North America. In Eocene times, however, the ocean broke through a breach in the Andean barrier and flooded an area east of the Andean highlands, forming a shallow epicontinental sea. The main Andean orogeny occurred during the Mid-Miocene (between 17 and 15 million years ago), since which time marine sedimentation has been confined to the extreme western margin of South America. The continental plains which now separate the old shields from the Andes are largely covered by clastic materials eroded from the great western cordilleras.

In the far north, a thick wedge of largely marine Cretaceous and younger sediments lie adjacent to the Caribbean orogenic belt. These have acted as reservoirs for hydrocarbons, giving rise to the very important oilfields which extend from Venezuela into Trinidad.

Aerial view of the Andes in Chile.

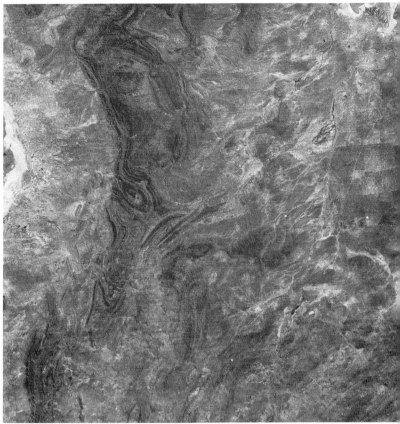

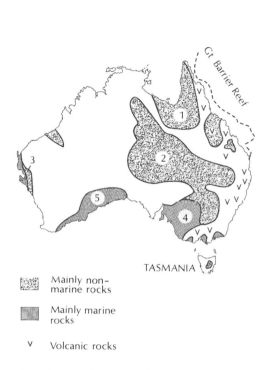

Mainly non–
marine rocks

Mainly marine
rocks

v   Volcanic rocks

Map showing the Tertiary basins of Australia:
(1) Carpentaria Basin, (2) Eyre Basin,
(3) Carnarvon Basin, (4) Murray Basin,
(5) Eucla Basin.

Landsat photograph of the Flinders Range, Australia. This uplifted
block of folded Proterozoic sedimentary rocks with infolded
Cambrian strata runs northward for hundreds of kilometres through
the southern half of the state of South Australia. It lies east of a
downfaulted trough (left-hand side of photo) known as the
Sunklands. The mountains attain a length of between 900 and
1200 m. Cenozoic sedimentary strata cover the regions to the west
and east of the range.

Australia The Tertiary history of Australia followed a similar pattern.
By Late Mesozoic times, the upheavals associated with the Tasman
orogenic belt were all but over and subsequently the craton was
warped and distorted but not folded, while there was very vigorous
igneous activity in some places. The large Murray Basin, which is
bounded by the Flinders and Mount Lofty Ranges, together with the
Carpentaria, Lake Eyre and Darling-Warrego Basins collected continental sediments that range in age from the Late Cretaceous to the
present. These include coal deposits. Marine deposits – mainly limestones – of Tertiary age are restricted to basins that developed along the
continental perimeter. During Palaeocene times, however, the sea
invaded these and reached its maximum extent during the Late
Eocene. Subsequently it retreated and continued to do so for the
remainder of the Tertiary period.

In contrast to the relatively stable conditions on the craton, there
was considerable crustal activity in the east; indeed these Eastern
Highlands have been recently active. Uplift of this tract began in Early
Tertiary times and peaked in the Miocene epoch; somewhat more

minor phases of uplift have continued into the Pleistocene. The Miocene disturbances were enough to drive the sea out of the marginal basins that flanked this region. Volcanicity occurred during Palaeocene and Eocene times, in a broad region along the eastern and southern border of the continent, including Victoria, Queensland and Tasmania. The volcanic rocks, which include alkali-basalts, nepheline-bearing lavas and trachytes, were almost certainly erupted when Australia and Antarctica began to drift away from one another about 53 million years ago.

New Zealand New Zealand was of low relief at the beginning of Tertiary times and was inundated by the sea during the Eocene and Oligocene. In later Oligocene times the sea retreated once more. Tertiary sedimentation was, therefore, dominated by marine clastic rocks, these being interbedded with some limestones and coals. Andesite lavas are abundant in the succession and the most acceptable explanation for their occurrence is that New Zealand was an active island arc throughout most of Tertiary time.

India While the Indian story is dominated by the rise of the Himalayas, there was another Tertiary event of great significance. During the Early Tertiary, an important phase of basalt eruption, centred on the Deccan region north of Bombay, gave rise to the 3000 m thick Deccan Basalts, a series of fluid plateau-type flows comparable with the Columbia River Basalts of North America. It is probable that most, if not all of these flows, were related to the separation of the Seychelles islands from peninsular India which commenced about 70 million years ago.

Substantial continental deposits worn from the Himalayas have been dumped onto the craton and have continued to accumulate just south of the mountain range since Mid-Miocene times. This has contributed to the infilling of the alluvial basins of the North Indian Plains and the building of the huge deltas of the Indus, Ganges and Brahmaputra. Elsewhere on the craton, continental deposits are thin. Marine deposition has been confined to basins situated on the continental margins, these collecting Tertiary sediments which, in the Cambay Basin, are between two and three thousand metres thick and are oil-bearing.

Antarctica Antarctica's recent history is, to say the least, sketchy; however, in recent years geologists have been probing this ice-covered continent with geophysical instruments and now have a far better idea of what lies beneath the ice than, say, fifteen years ago. In general terms the geological story is not dissimilar to that of South America. Tertiary sedimentary and volcanic rocks accumulated in basins confined to the Pacific margin, between South Victoria Land and the Palmer Peninsula. The volcanic activity and the orogenesis associated with this were probably related to the eastward movement of oceanic lithosphere beneath the continent.

Life and Death

During Jurassic and Early Cretaceous times the terrestrial landscape would have been dominated by reptiles and seed-bearing plants (gymnosperms). Marine environments were the home for molluscs, ammonites, fishes, echinoids, crinoids and a variety of other shelly forms. Crustaceans, also, lived in the nearshore waters. Later in the

Cretaceous, the more advanced angiosperms joined the plant kingdom and have persisted to the present. Among their flowers and branches fluttered the first butterflies.

There is no doubt that at the close of Mesozoic times, life was very diverse. Not all organisms enjoyed conditions on Earth, however, as at the close of the Mesozoic a large number of important animal groups became extinct in response to a change in conditions. Most famous of these groups were, of course, the dinosaurs; but many other reptiles died out, although the crocodiles, lizards, turtles and snakes survived. Many marine invertebrates also disappeared before the Tertiary period opened, the ammonites being a case in point; others, however, survived apparently unaffected. Many theories have been put forward to explain these mass extinctions, but no single one appears to have been universally accepted.

The Death of the Dinosaurs

Evidently a great change affected the surface of Earth about 65 million years ago. The dinosaurs, which had been unchallenged for so long, died out, leaving the stage set for the evolution of the more agile and intelligent mammals.

The extinction of the dinosaurs occurred very suddenly on the geological time scale. Of course, this is not to say that it happened overnight, but even so it was very rapid considering that the dinosaurs had been anything but an unsuccessful biological experiment. They ruled the world for a great deal longer than Mankind has done up to the present time.

What happened? Bearing in mind that it was not only this group that suffered the extinctions, there must undoubtedly have been a fairly abrupt change in their environment. One currently popular theory is that the Earth was hit by an asteroid several miles in diameter. This is by no means out of the question. Although the largest members of the asteroid swarm keep strictly to that region of the solar system between the orbits of Mars and Jupiter, there are many smaller ones that swing inward and make close approaches to the Earth, so that occasional collisions must be regarded as inevitable. An asteroid more than a dozen miles in diameter would cause tremendous damage, and the effects would be global rather than localized. Climatic changes might follow, and it has been suggested that the dinosaurs, with their cumbersome bodies and small brain power, were unable to adapt to new conditions. Some support for the collision theory can be drawn from the fact that some rocks laid down 65 million years ago contain more than normal amounts of iridium, an element that is rare on Earth but which could well have been augmented from outside by an asteroid.

Alternatively, a slower climatic change, not due to an asteroid, may have proved fatal to the dinosaurs, either because their food supplies ran short or because the shallow seas retreated, depriving them of their most favoured environment. If this is true, then presumably the cause lies in some slight change in the output of the Sun. Then, of course, there may have been new competition for their favoured food, as new species of birds and mammals evolved, or the swampy habitats the herbivorous varieties inhabited may have been dried up during the general rise in land level at the close of the Cretaceous. There are numerous ways it may have happened. But nobody really knows; all we can say with certainty is that at the end of the Cretaceous, the great reptiles made their exit.

21

NEW OCEANS
FOR OLD

The Modern Oceans

The break-up of Gondwanaland caused dramatic changes in the configuration of land and sea. Before the Mesozoic plate movements, the Pacific and Tethys were the Earth's principal oceans. Of these two, the latter separated Laurasia from Gondwanaland, and has since been virtually eliminated, the only surviving remnants being found in the Black and Caspian Seas. The Pacific, on the other hand, still remains the largest ocean, although it is slowly decreasing in area, largely due to subduction along its western margins. Although it is floored predominantly by rocks that are 200 million years old or less, there has been a Pacific Ocean to the west of the Americas for much longer. Certainly it is not as old as the Moon, which was once thought to have been spawned from it, but it dates back to the Early Palaeozoic, and possibly may be even older.

In contrast to these two ancient oceans, the Atlantic, Indian and Antarctic oceans are all relatively young. They are the new oceans that grew as Gondwanaland broke up and its components drifted slowly apart, eventually giving us the modern pattern of land and sea. Oceanic growth was made possible by the production of new oceanic crust at mid-oceanic ridges. The Mid-Atlantic and Carlsberg Ridges are two of these (the latter is situated within the Indian Ocean). This new oceanic crust is added to their margins as the gulf between the adjacent continental fragments increases.

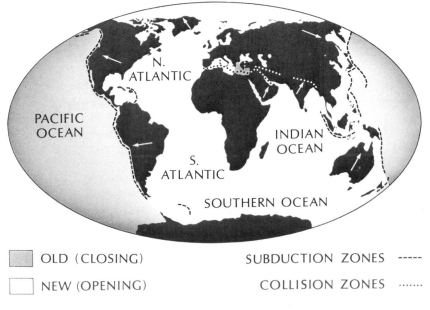

The opening and closing oceans.

One significant difference between these new oceans and the Pacific is their plate tectonic situation. One of the principal features of this great ocean is that it is largely encircled by subduction zones; thus the oceanic crust is not structurally coupled with the adjacent continents. This means that although active spreading is occurring oceanic lithosphere can also be continually got rid of down marginal subduction zones, and the area of the ocean actually diminishes. With the younger oceans like the Atlantic, however, things are quite different; here the oceanic lithospheric plates are composite, that is, they consist of both oceanic and continental crust which are converging. There is no destructive subduction at the Atlantic Ocean margins, so the ocean continues to grow. Thus we entitle this chapter 'New Oceans for Old' – because the young ones expanded in area at the expense of their older counterparts.

The Mid-Atlantic Ridge

The Mid-Atlantic Ridge runs as a continuous mountain range down the axis of the ocean, marking the line along which the African, North and South American and Eurasian plates are currently separating. The crest of the ridge is ofest many times by fractures which run roughly at right angles to it; these are the fracture zones, which are inactive, and the active transform faults which we mentioned earlier; these are often seismically active, particularly where they intersect the ridge. It is odd that the Mid-Atlantic Ridge is at once the oldest and youngest part of the Atlantic: while it is geographically old it is relatively young in geological terms.

The simplest part of the ridge is the central portion, which parallels the curves of the African and American continental margins. To the south of this it is offset by a number of fracture zones. South of the Equator it becomes straighter and eventually turns east around the tip of Africa, eventually linking with the Indian Ocean ridge. To the north, on the other hand, things get rather more complicated. Thus, the ridge passes through the Azores and separates Newfoundland from Spain. Between Labrador and along Ireland it is offset along the Charlie-Gibbs fracture zone. It then strikes northeast to form the Reykjanes Ridge which runs into Iceland and was the first mid-ocean ridge segment at which symmetrical magnetic lineations were understood in terms of sea-floor spreading processes.

Iceland is the largest land region made from oceanic crust alone. There is also a negative gravity anomaly here which strongly suggests that mantle material must lie at unusually shallow depth. It is built from plateau basalts and lavas extruded from central volcanoes; there are also thousands of dikes, mainly of basalt. Generally speaking the lavas get older with increasing distance from the island's axis – away from the active spreading centre. The oldest lavas so far dated by the radiometric method yield ages of about 16 million years, which is very much younger than the lavas of east Greenland and the Faeroes (60 million years).

Iceland itself is a part of the ridge and is also part of a transverse topographic rise which joins the ridge to Greenland and the Faeroes, roughly along the Arctic Circle. During Tertiary times a tremendous amount of igneous activity took place here, giving rise to the extensive plateau basalts of the Faeroes, east Greenland and northwest Ireland. North of Iceland, the ridge continues northwards to the volcanic island of Jan Mayen, then is offset to the east. It passes quite close to the west

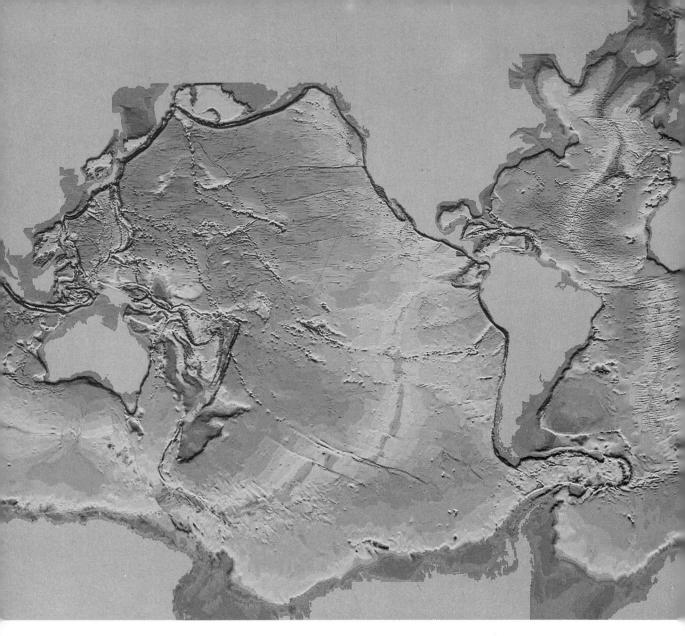

coast of Spitsbergen, then, via a series of offsetting transform faults, into the eastern part of the Arctic Ocean. There it can be traced beneath the ice of this ocean until it apparently dies out near the Lena delta of the Asian continent.

The Red Sea and the Gulf of Aden

The Red Sea and Gulf of Aden are of particular interest in the context of young oceans which are in the process of spreading. The axis of the Red Sea is an active spreading centre: the entire region is characterized by high heat flow, recent continental volcanic activity and current seismic disturbances. Calculations suggest that rifting started in the Late Tertiary and that for the last 10 million years, sea-floor spreading has been occurring at the rate of about 0.4 cm per year. The local presence of extensive basalts of Late Cretaceous-Early Miocene age and the presence of very thick evaporites beneath the waters of the Red Sea and along its perimeter are both features we have already noted as being characteristic of the early stages of continental rifting.

The oceanic Carlsberg Ridge is offset into the Gulf of Aden from the

ABOVE Bathymetric map of the world's oceans as acquired by the SEASAT satellite. Computerized contouring beautifully brings out the main submarine structures, including rises and transform fractures.

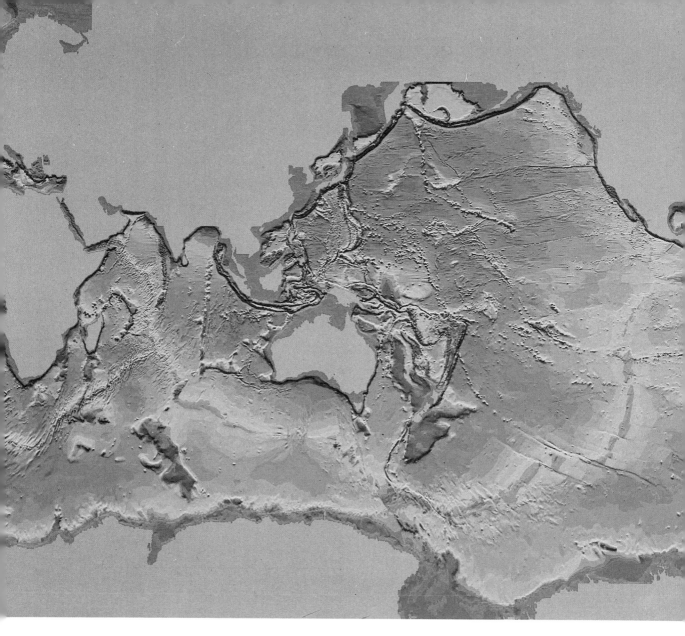

RIGHT Volcanic activity in Iceland: Víti from Krafla.

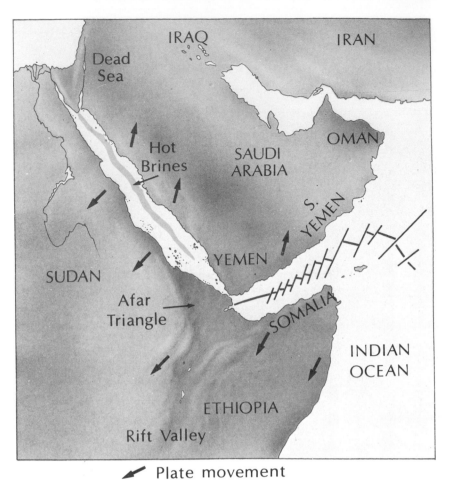

Map of the Red Sea and Gulf of Aden. Alfred Wegener observed that if the Afar triangle is considered to be a recent addition, then the two sides of the sea fit quite snugly together. The regions of hot brines described on page 131 are also shown.

⬋ Plate movement

east, while to the west there is a direct connection with the continental Rift Valley of East Africa. At the northern end of the Rift, in Ethiopia, the active extensional systems of the Red Sea, Gulf of Aden and East African Rift Valley meet in area known as the Afar Triangle. This is a low-lying plain which is underlain by a thick sequence of Tertiary and younger sedimentary rocks. It appears to have been domed up during Early Tertiary times and thence became the focus for this very recent continental rifting.

The Indian and Southern Oceans

At once separating yet linking the younger Atlantic Ocean and the more ancient Pacific are the Indian and Southern Oceans. From the deep ocean floor rise oceanic ridges that delineate an inverted 'Y' and define the boundaries of three diverging lithospheric plates. The Carlsberg Ridge runs north-south between peninsular India and the east coast of Africa and is diverted into the Gulf of Aden at its northern end by a series of transform faults. To the south this ridge forks. The westerly fork trends southwest, then almost due west along the ocean floor between Africa and Antarctica, eventually joining the Mid-Atlantic Ridge at its extreme southern end. The easterly branch strikes in a direction roughly east-southeast, between Australia and Antarctica. Palaeomagnetic traverses made across the ridge to the south of Australia indicate a spreading rate of about 3.5 cm per year on either side. Using this information and correlating magnetic stripes here with those in the Pacific Ocean, it is clear that the ocean separating Australia from Antarctica has spread for only 35 million years.

Deep-sea coring and seismic studies show that very large areas of the ocean bed on either side of the mid-ocean ridges have little or no sedimentary cover; this suggests that the ocean floor is very youthful. The thickest accumulations occur in the Southern Ocean where 'oozes' made from siliceous organisms (diatoms) give rise to sequences which may be anything from 100 to about 750 m thick. Land-derived debris is most evident on the Indian Ocean floor off the mouths of the great Ganges and Indus rivers, where vast submarine fans have been under construction ever since the Himalayas began to rise. The average rate at which this riverborne material has accumulated is of the order of 17 cm per 1000 years. Most of the debris is well stratified and consists of turbidites. Other substantial accumulations are found off the east coast of Africa, where thick sediments have accumulated in the Somali and Malagasy Basins. Madagascar itself is a fragment of continental crust surrounded by oceanic crust. The very old gneisses that form the backbone to the island are overlain by Late Palaeozoic and Mesozoic rocks, which collected in fault-defined basins. The Seychelles also have continental characteristics, being largely built from Early Palaeozoic granites, and are believed to have formed part of the continent of Gondwanaland before being rifted away during its fragmentation.

On the northeast side of the Indian Ocean is an island arc system. This emerges as the island arcs of Sumatra and Java. To the south and west of these are deep trenches and sediment-laden troughs that are associated with northward-dipping Benioff Zones. It is along this line that the northward-moving Indo-Australian Plate is in contact with the Eurasian Plate. The extreme distortion and crumpling of the island arc system is a direct result of tectonic interaction between the two plates.

Map of the Indian Ocean showing oceanic ridges and transform faults. The boundaries of the main crustal plates are shown in inset.

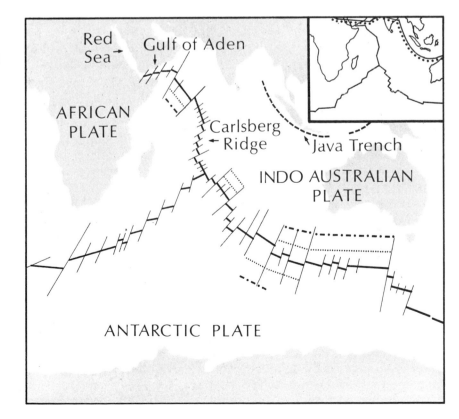

177

The Ocean's Margins

We have seen how, before the break-up of Gondwanaland, faults developed in the continental crust and that there was also increased igneous activity preceeding rifting, giving rise to vast tracts of plateau basalts and also intrusive igneous rocks close to the continental margins. As the crust was stretched by plate motion, the faulting and related subsidence caused marginal basins to form – these now residing along the margins of the modern continents. We recognize the same events in the history of the Atlantic Ocean. In the early stages, when the new ocean was little more than a shallow and very restricted trough, sea water would have flooded in, only to be evaporated quickly due to restricted circulation and shallow depth. Therefore the first marginal deposits were the Triassic salt deposits mentioned earlier.

As the ocean widened, so more normal marine shelf deposition developed and, as the thickness of the sediments increased, so the continental margins became more depressed. Where large river systems flushed huge amounts of debris out onto the ocean floor, as for instance was the case with the Amazon and Niger Rivers, massive deltas also aided in the building work. Where deep boreholes have been sunk along the seaboard of eastern North America, they show that sedimentation began as soon as continental separation started, sedimentary sequences of 3 km or more being quite usual. Often deep canyons incise the shelf regions, and debris dumped near to the edge of the shelf may be flushed down to deeper levels by turbidity currents. These were first documented off eastern Newfoundland, where they have frequently broken telegraph cables.

On the eastern margin, the Atlantic also has a number of large basins

Tengger massif, Java, showing volcanic cones of the Indian-Pacific ocean island arcs.

178

Nimbus-7 coastal zone colour scanner image of the Western Approaches of the British Isles; taken on 4 April 1980. The yellow and brown colours indicate high levels of sediment in the surface waters of the continental margin. The blue colours may be due to plankton 'bloom'.

which are today receiving marine sediments. The North Sea Basin is in reality a 'failed ocean', and is atypical. However, to give some idea of how much sediment can accumulate in a fairly young basin, off the coast of Holland, the infilling Mesozoic and younger sediments are 6 km thick and are pierced by salt domes which act as very efficient traps for the hydrocarbons which have collected there. Other basins lie off the Western Approaches to the English Channel, and there are also the Aquitaine and Lusitanian Basins, further south.

The western margin, however, is complicated slightly by the presence of short island arcs at two isolated locations along the otherwise passive continent-ocean boundary. The Lesser Antilles form an active island arc, separating the Atlantic from the Caribbean, a separate oceanic basin. In the far south is another active zone – the Scotia Arc which links the active orogenic belts of the Andes and the Antarctic Peninsula. These two small regions of subduction are exceptional, however, and the Atlantic margins are fundamentally of passive type.

Finally, one feature of the modern Atlantic perimeter – namely the Iberian Peninsula – very effectively spoils the otherwise good 'fit' between America and Eurasia used in attempts to reconstruct the pre-drift configuration of the two continents. Its position, and the existence of the Bay of Biscay, are explained by the palaeomagnetic data, which reveals that Eocene lavas in Spain have a rather different magnetic orientation than similar rocks from elsewhere on the continent of Europe. However, by rotating the Iberian Peninsula about 40 degrees in a clockwise direction, these orientations can be brought into line. This makes it clear that the peninsula rotated 40 degrees anti-clockwise, away from France, during the opening of the Atlantic.

22

THE PACIFIC OCEAN

Pacific: An Ancient Ocean

Although the oldest oceanic crust on the floor of this great basin is little more than 200 million years old, the feature itself is much more ancient and its history spans the whole of Phanerozoic time and probably extends even further back. Among the most fundamental differences between it and the Atlantic Ocean is, as we have seen, its virtual encirclement by active subduction zones. These have acted out a major role in geological history as they have made it possible for a balance to be maintained between the creation of oceanic crust at spreading centres and its loss at destructive margins.

Unlike the other oceans, the Pacific has its oceanic ridges, but in this case they are not symmetrically placed. The main ridge enters the basin from a point between the southern tip of New Zealand and Antarctica, then turns northward at about longitude 120 degrees west, and runs approximately north-south as the East Pacific Rise. This great submarine range is among the most closely studied parts of the terrestrial oceans, and has been traversed by oceanographic vessels using the most up to date geophysical devices and inspected at close quarters by men enclosed in specially constructed craft called submersibles.

As is the case with the Mid-Atlantic Ridge, the ridge is offset by numerous transform faults. Several of these connect the East Pacific Rise with an active spreading centre in the Gulf of California. From then on, apart from minor spreading centres, there is no active sea-floor spreading offshore of the western U.S.A. The distinct asymmetry of the ocean about this ridge is due to the fact that the oceanic crust to the east has been overridden by the North American continent. Further south, the oceanic plate to the east of the Rise is being subducted beneath South America. Such a situation is not found with the Atlantic or Indian Oceans.

Oceanic Ridges and Rises

Very refined techniques devised by commercial companies interested in the rich sulphide deposits localized along the East Pacific Rise have increased our knowledge of the Pacific manyfold during the past few years. Furthermore, numerous very recent dives by the French submersible *Cyana* have elucidated some of the details of how new oceanic crust is formed. Thus by profiling to and fro across the East Pacific Rise it has been possible to show that it is a much broader feature than the Mid-Atlantic Ridge. Atop the crest is a ridge roughly 30 km in width and 500 m high. It is cleft by a rift along its axis and descends gently on either side to the abyssal depths. The greater breadth of the Rise is almost certainly a function of the rate at which sea-floor spreading is occurring here; study of the magnetic striping reveals different spreading rates at different points, but shows the average rate to be 4.5 cm per year. This is substantially greater than in

An artist's impression of lava domes along a mid-oceanic ridge.

the Atlantic. Using this rate as a basis for calculation, only 40 million years would be necessary for the whole of the eastern Pacific sea floor to have formed.

Along the northern portion of the Rise, west of Central America, its structure is rather complicated. Numerous major transform faults offset it, and these are spaced about 200 to 300 km apart. Between the principal faults are many smaller ones; these may be as little as 10 km apart. Close study of these during the last few years has shown that near the major transform faults the Rise is quite deep; it rises to shallower depths more or less midway between each pair of faults. Thus between the major fractures the profile is that of a broad lava dome. Smaller domes - they look more like large 'blisters' — occur between the smaller fractures, and it appears that each of these may be

Hydrothermal vents ('black smokers') along the East Pacific Rise. The dark coloration is due to the expulsion of sulphides that have been dissolved out of the underlying rocks.

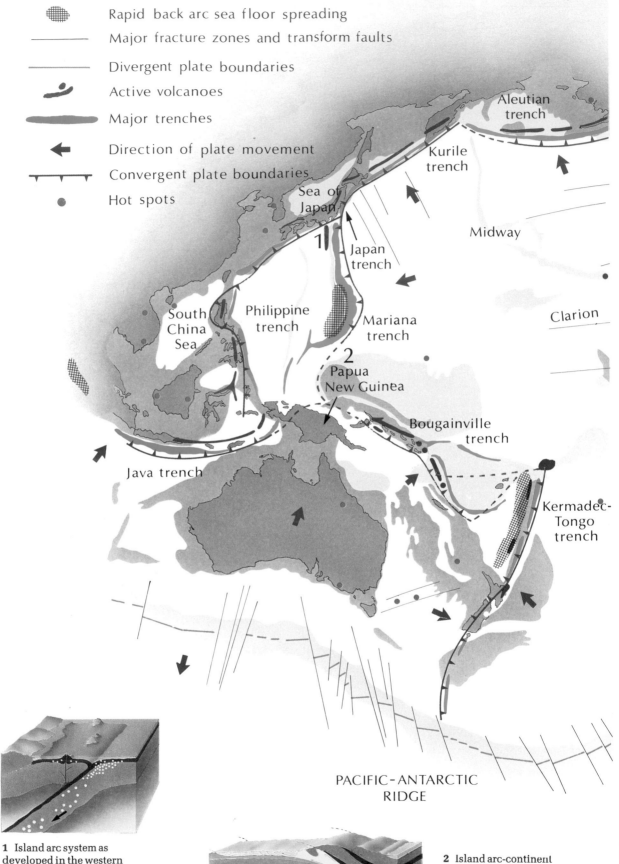

Rapid back arc sea floor spreading

Major fracture zones and transform faults

Divergent plate boundaries

Active volcanoes

Major trenches

Direction of plate movement

Convergent plate boundaries

Hot spots

Aleutian trench

Kurile trench

Sea of Japan

Japan trench

Midway

South China Sea

Philippine trench

Mariana trench

Clarion

Papua New Guinea

Bougainville trench

Java trench

Kermadec-Tongo trench

PACIFIC-ANTARCTIC RIDGE

1 Island arc system as developed in the western Pacific. Earthquake foci shown as white dots.

2 Island arc-continent collision as seen in the region of Papua New Guinea.

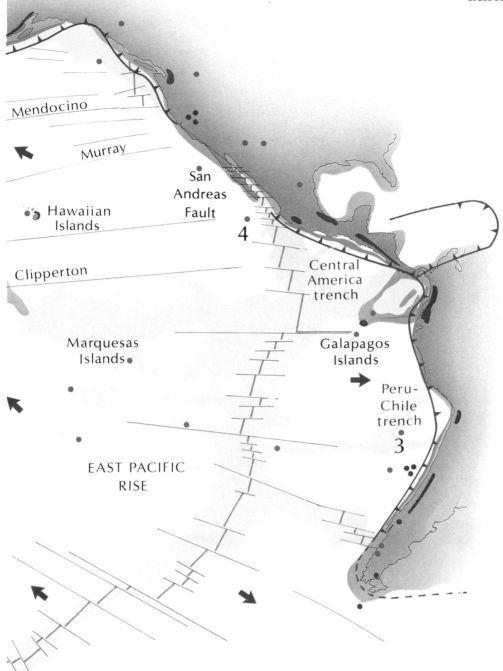

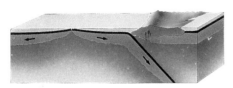

Mendocino

Murray

San
Andreas
Fault

4

Hawaiian
Islands

Clipperton

Central
America
trench

Marquesas
Islands

Galapagos
Islands

Peru-
Chile
trench

3

EAST PACIFIC
RISE

3 LEFT Ocean-continent style of subduction as found along Andean coast of South America.

4 RIGHT Transform faults offsetting segments of oceanic ridge.

a miniature spreading centre. It has been suggested, therefore, that instead of there being just a few major spreading centres, the new oceanic crust is generated at a much larger number of smaller ones. This was unsuspected a few years ago.

What of the structure beneath the Rise? Study of seismic waves shows that the East Pacific Rise has a deep 'root' which goes down at least 200 km. The Low-Velocity zone, where seismic waves travel at their minimum velocity, is normally at roughly half this depth. Seismic data reveals that beneath it are large magma chambers of gabbroic composition; sometimes blocks of coarsely crystallized igneous rocks are carried up to the surface in the basalt flows that are extruded there.

To try and learn more about the subsurface structure, in 1981 a deep hole was drilled into the ocean floor in a region known as the Costa Rica Rift. This crosses the Rise between South America and the volcanically active Galapagos islands. The bore hole, known as 504-B, penetrated the sea floor to a depth of 1076 m and, although this does not sound particularly impressive, was a major achievement in view of the most difficult conditions encountered in deep-ocean drilling. The crust hereabouts is known to be 6 million years in age. The core reveals that the ocean floor here is covered by 275 m of recent sediment; beneath this, the next 575 m is made from pillow lavas and brecciated volcanic rocks of basaltic composition. Between depths of 575 and 780 m, dikes appear within the succession then, at greater depths, pillow lavas and brecciated rocks give way to massive basalts and abundant dikes. This was the first proof we had of exactly what does lie beneath the Pacific Ocean floor.

Island Arcs of the Pacific

Arcuate chains of islands festoon the northern and western margins of the Pacific; they also extend southwards through the Philippines to New Caledonia and New Zealand. These island arcs not only have their curved form in common, they all have deep trenches which extend down to 10,000 m in some cases, and lie on the ocean side of the arcs. They are also zones of active volcanicity and seismicity, the seismic activity being along steeply dipping surfaces that dip away from the ocean. These 'Benioff Zones' are the surfaces along which oceanic lithosphere is subducted. The earthquake activity and volcanicity associated with these zones is a direct result of plate movements. Subducted oceanic crust returns to the mantle on reaching the arc-trench systems and is subsequently recycled within the Earth.

Of all the world's island arcs, the Japanese islands have been most intensively studied, and have played an important role in the development of modern ideas about plate tectonics. At first sight the islands might be considered to form four different arcs, but when the submarine topography is carefully studied, we see that there are, in fact, only two: one joins the Kurile, northeast Honshu and the Izu-Bonin-Marianas island groups (East Japan Arc), while the second links the Ryukyus, Kyushu, Shikoku and western Honshu groups (West Japan Arc). On the Pacific-facing flank of each arc are sited the deepest trenches. Asia and Australia sit well back from the active arcs, and between the island groups and the continental margins are well-developed marginal seas, like the Sea of Japan.

There are marked differences between northeast and southwest Japan. Thus while the former shows features that are typical of active arcs (volcanicity, deep trench) the latter is volcanically quiet. The

The arrangement of metamorphic belts in Japan. The Median Tectonic Line and Fossa Magna are shown by heavy lines. High-temperature/ low-temperature belts (dotted) and low-temperature/high-pressure belts (stippled) are also shown.

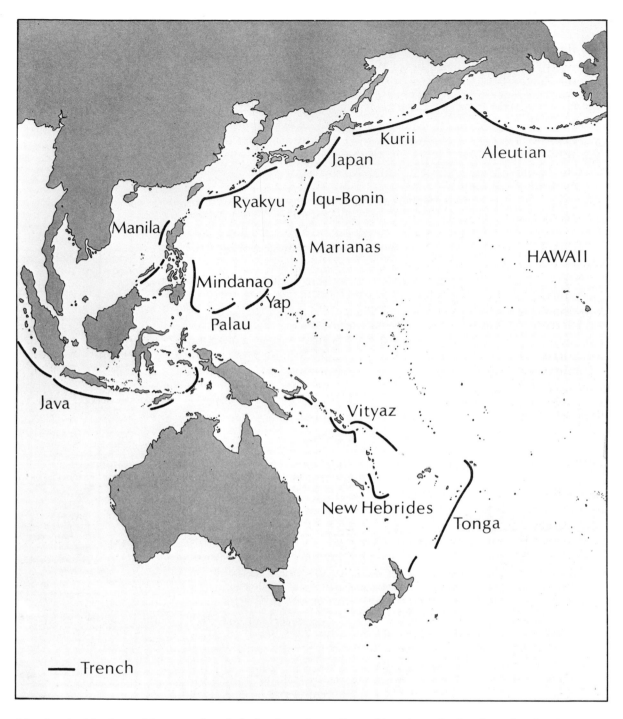

Map showing island arcs of the western Pacific.

volcanic belt of northeast Japan lies along the landward side of a major discontinuity called the Fossa Magna, which it turns into, then extends southwards into the Izu-Bonin-Marianas arc. Southwest Japan, on the other hand, has a 'grain' that trends roughly parallel to the island's axis, and includes metamorphic belts which occur in pairs.

Situated slightly landward of the deepest trenches, there are well-defined negative gravity anomalies, implying a mass deficit. The obvious conclusion to be drawn is that rock of relatively low density is present beneath the island arc, and that this rock is oceanic crust of the Pacific plate, which is descending into, and displacing, the denser rocks of the mantle.

185

The East Pacific Margin and San Andreas Fault

The eastern margin of the Pacific is very different from the western, for here the Pacific has lost ground to the westward-moving plates that carry the Americas and the character of subduction is very different. In the Gulf of California and northwards from here, the boundary between the Pacific and North American plates is offset by numerous small transform faults. Earthquake foci are typically quite shallow (10 km) and are concentrated both along the boundary and the transverse fractures which offset it. Along this margin of the Pacific is the famous San Andreas Fault.

Although the individual movements along the San Andreas Fault, and others like it, are quite small, the cumulative effect of such movements over a period of millions of years is considerable. Thus while the amount of horizontal displacement experienced along the fault during the famous 1906 San Francisco earthquake was only 5 m, and the vertical movement accomplished by the earlier earthquake in Owens Valley 4 m, if such movements were repeated many times over, substantial vertical and horizontal displacements of the crust could be accomplished in a million years.

During Jurassic and Tertiary times the North American continent overrode the eastern floor of the Pacific, and there is currently no deep trench nor are deep-focus earthquakes experienced along this stretch of the ocean's coastline. Today one plate, which comprises the great majority of the North Pacific Ocean together with a slice of western California, is in contact with the North American plate which includes the remainder of North America and the western part of the Atlantic Ocean. The San Andreas Fault separates the two along much of the western coast of the U.S.A. Because the Pacific plate moves inexorably northwestwards with respect to North America, and towards the Aleutian trench, so California experiences frequent earthquakes in response to these subcrustal movements.

The 'East Pacific' (Farallon) plate – which has now largely been subducted beneath North America – nevertheless still leaves two remnants behind; one lies south of the San Andreas Fault and is called the Cocos plate, the other is a smaller fragment which intercepts the American coastline between northern California and Vancouver (Juan de Fuca plate). Northeasterly or easterly subduction is still going on along those stretches of the western coastline where these remnants abut against it.

As the East Pacific Rise was gradually overrun by the North American plate, so an unstable 'triple junction' developed where the three plates (the existing Pacific plate, the largely overrun 'East Pacific' plate now represented by the Cocos remnant, and the North American plate) met. This triple junction slowly migrated along the continental margin as more and more of the Pacific plate came into contact with North America. An active triple junction exists further south, where the Pacific, Cocos and Nazca plates meet.

The situation further south is quite different again; here there are deep trenches off the western coasts of Central and South America; furthermore Benioff Zones dip steeply under the South American continent. Branches of the East Pacific Rise trend eastwards towards Central America and also towards southern Chile, these defining the smaller Nazca Plate that is currently plunging beneath the South American plate. The story of the Andean Cordillera, formed in direct response to these plate movements, is described in the next chapter.

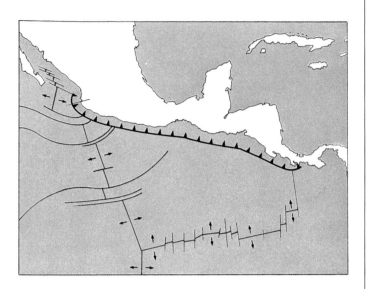

ABOVE Plate boundaries to the west of Central America, showing the junction between the Cocos Plate (centre), Pacific Plate (left) and Nazca Plate (bottom right). Spreading axes shown in red, convergent plate boundaries by toothed lines.

RIGHT The convergence zone between the Pacific, Juan de Fuca and North American plates.

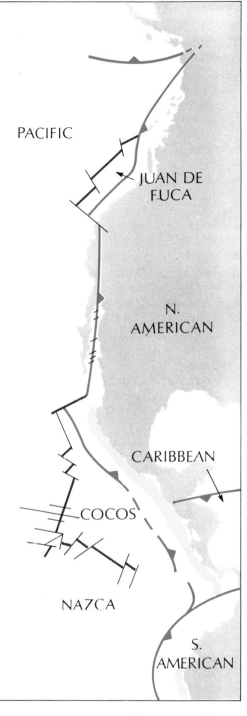

PACIFIC

JUAN DE FUCA

N. AMERICAN

CARIBBEAN

COCOS

NAZCA

S. AMERICAN

— Constructive plate margins

▲ Destructive plate margins

— Transform faults and fault zones

Oceanic trenches

23

THE CORDILLERAN CHAINS

Mountains of the Western Americas

We have already followed the Precambrian and Palaeozoic history of the North and South American cratons; thus far we have not looked in detail at the geological history of their western margins. In a sense the events that shaped these mighty mountain ranges were similar to those that fashioned the Appalachians and the Caledonides, but whereas volcanic and tectonic activity is now minimal in these regions, the American Cordilleras are very much active to this day – witness the eruptions of Mount St Helens and of El Chichon!

Together the Cordilleras of North and South America form an elevated mountain chain that runs from Alaska to Mexico, and from Guatemala to the southern tip of Chile: a distance of almost 18,000 km. Situated as they are on the margin of these continents, they define a belt that has remained geologically active throughout the Phanerozoic. The two continents have not always occupied the same relative positions; the modern disposition is a result of relatively recent movements since the break-up of Pangaea. In Palaeozoic and probably earlier times they formed part of two different supercontinents. Since the Atlantic first started to open, about 200 million years ago, both North and South America moved westward on the leading edges of plates that grew from the Mid-Atlantic Ridge. It is likely that both continents, however, have been overriding Pacific Ocean crust since Palaeozoic times, although no collision with other major continental plates ever occurred. The present mountains are the latest of a succession of older chains; their great elevation is due to recent movements.

The Northern Cordilleras

The growth of the western cordilleras of North America has been largely a function of plate movements involving the Pacific, Juan de Fuca (Farallon) and Cocos plates. The details, which have been worked out after the study of magnetic anomaly patterns, radiometric dates and geochemical data, are still a matter for argument but the general picture has now become clear. Thus, whereas prior to Tertiary time the Farallon plate had been moving northeastwards, the Pacific plate was moving northwestwards; when more and more of the Pacific plate gradually came into contact with the advancing North American plate, about 30 million years ago, an entirely new tectonic situation arose. Many of the geological features of this region seem to have formed due to the distortion of western North America in response to sideways slippage along the contact between the (now) almost totally

The North American Rockies: British Columbia.

Landsat mosaic of part of Oregon and Washington states, as prepared by
General Electric Photographic Laboratory, Maryland. This fine mosaic
shows the western seaboard of America from the Fraser River delta
(top of photo) to just south of Klamath Lake (near bottom of picture). The
Columbia River runs along the state boundary, approximately midway up
the frame. Very prominent are the western cordilleras, from Mount Baker
in the north, through Mount St Helens (north of the Columbia River), to
Crater Lake – an old volcanic caldera (the circular lake north of Upper
Klamath Lake). The volcanic mountains along this chain have arisen
above the contact between the Juan de Fuca and North American plates.

subducted Farallon plate, the Pacific plate and the North American
plate. Such lateral motion would not only distort the continental crust
but would also be capable of displacing it over distances of thousands
of kilometres.

Before the Late Cretaceous (about 80 million years ago), large-scale
volcanicity was confined to California and Nevada. Subsequently it
migrated eastwards until, between 70 and 65 million years ago, it had
moved into the central regions of north Nevada and the Rockies. Then,
at the beginning of Tertiary times, the mountains of Colorado were
elevated during the Laramide orogeny. The 2-kilometre vertical uplift
of the Colorado Plateau, which was accomplished without distortion of

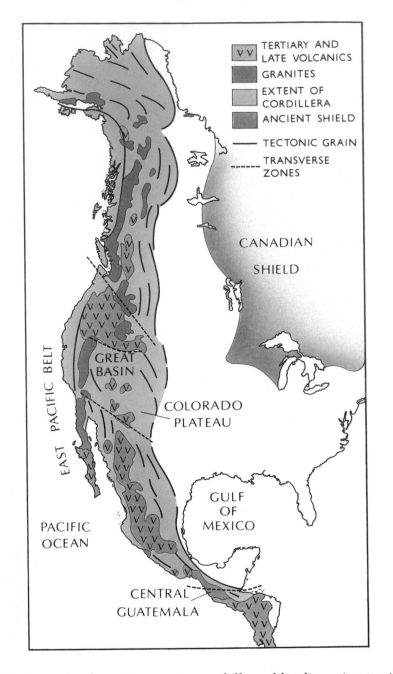

Legend (map):
- V V TERTIARY AND LATE VOLCANICS
- GRANITES
- EXTENT OF CORDILLERA
- ANCIENT SHIELD
- —— TECTONIC GRAIN
- - - - TRANSVERSE ZONES

CANADIAN

SHIELD

EAST PACIFIC BELT

GREAT BASIN

COLORADO PLATEAU

GULF OF MEXICO

PACIFIC OCEAN

CENTRAL GUATEMALA

its horizontal sedimentary strata, was followed by dissection to give dramatic features like the Grand Canyon. The central parts of the Rockies were by now rising and, during Oligocene time, 135 million years ago, great outpourings of andesitic lavas flowed over much of southwest Colorado. Then came the major change in plate movements. Large numbers of fissures opened, as the basin-and-range province was stretched by a factor of two. The extensive Columbia River basalts were extruded during this phase. Finally, western central North America was uplifted about 2 km over an area of 335,000 km^2 to give the modern Rocky Mountains.

Calculations suggest that at least a 7000 km wide strip of oceanic crust must have disappeared down the great trenches under western North America as the Farallon plate was subducted. Many of the vast granite batholiths of the Cordilleras, such as the Coast Range batholith of British Columbia, were undoubtedly produced during this subduction stage. However, while the Coast Range and Sierras may be related to oceanic subduction, the Rockies clearly are not. They are sited far inland, remote from the line where oceanic subduction presumably occurred.

Geochemical data from the volcanic rocks of the region shows that the plate from which the volcanic rocks of the Coast Range rose was at a level 100 km down when generated. Further east, under the high Sierras, it was twice as deep then, further inland again, beneath the border of Utah and Nevada, three times as deep. This conforms to the typical pattern of a descending oceanic plate moving eastwards. Surprisingly, however, further inland the lavas apparently were generated from an oceanic plate at much shallower depth (i.e. ~200 km).

The implication of this pattern for the geological history of the region is that there may have been not one, but two subducted plates involved. One explanation could be that subduction of the ancient Farallon plate more or less ceased when its leading edge had descended to a depth of 700 km inside the Earth and become embedded in unyielding cold mantle rocks. It then broke off beneath North America, so that its shallowest end lay along the line of what is now the Wasatch Range, a zone of continued instability. Subsequently subduction would have started elsewhere. The uplift of the Rockies themselves and the volcanicity that accompanied this would, if this scenario is valid, be due to the heating up and expansion of the buried oceanic plate underneath this region, in response to the elevated temperature of the surrounding mantle.

Finally, when subduction of the Farallon plate beneath North America had ceased, generation of the andesitic lavas typically produced along descending plates was halted. This effect migrated continually southwards as more and more of the Pacific plate came against North America. Gradually, as the North American plate felt less and less pressure from the Farallon plate, the crust became stretched and voluminous basaltic eruptions emerged. These slowly migrated towards the northwest, producing in their path the Columbia River plateau basalts.

The Southern Cordilleras

Linking the Americas is the archipelago of Central America which resides on the western edge of the North American plate. Off its western coast are deep trenches that are associated with subduction of

Section through the Andes showing the general structure of the region.

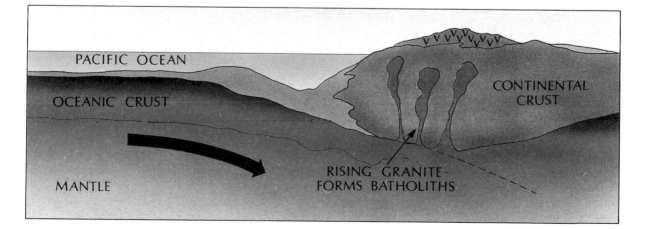

OPPOSITE Cenozoic life.

the Cocos plate beneath the archipelago. The present-day topographic expression of the Cordilleras of South America is the Andean chain – one of the most magnificent of all Earth's mountain ranges. The Andes themselves were raised up quite recently, but it should not be forgotten that orogenic disturbances affected the region as far back as Palaeozoic times.

The South American plate, which has South America on its leading edge, is currently advancing westwards and overriding the oceanic Nazca plate to the east. The latter is virtually stationary with respect to the Earth's deep interior. Along the plate boundary there are deep trenches which run two-thirds of the way down the continent and to the east, high cordilleras run parallel to the Chilean coast. A Benioff zone slopes at 25 degrees eastwards under the continent; this is the site of frequent and often substantial earthquakes. Approximately 300 km east of the trench line, the cordilleras begin to rise along the western continental margin. Active volcanicity is a feature of the Andean ranges, this being generated because the subducted oceanic slab has attained sufficient depth along this belt for melting to occur.

Calculations suggest that the Nazca plate may be descending beneath South America at the extraordinary rate of 10 cm per year! If this is the case, then thousands of kilometres of oceanic crust must have been subducted beneath South America during the current phase of plate movement. Particularly confusing in this context is the fact that relatively close to the coast of southern Peru, some 300 km west of what geologists believe to have been the earlier margin of the continent, are continental deposits which are at least 400 million years old (early Devonian). Some workers have suggested that this strip of coast may be the remains of a very small continent that originally lay off the shore of South America but was crushed against it as more and more of the oceanic plate was driven east.

In the more central regions of the Andes, particularly in Bolivia and Peru, two ranges run parallel to the continental margin. The rocks of the inland range, or Eastern Cordilleras, formed in the Palaeozoic in part from marine sediments that had been driven eastwards, away from the Pacific. The Western Cordilleras, on the other hand, are of Mesozoic and Tertiary age and are comprised of two belts; the more westerly is built from trench sediments, while that to the east from mainly shallow water debris. All of the rocks are highly deformed.

During the Late Mesozoic orogeny, enormous batholiths of granitic rocks rose into the mountain belt and widespread volcanic activity began. Cotopaxi, situated in Ecuador, is the highest active volcano on land, having an altitude above sea level of 5943 m. The last major phase of elevation – which gave rise to the modern aspect of the Andes – appears to date from the Late Neocene (10 million years ago). It did not involve any further deformation, but seismic and volcanic activity continues to this day.

The southern part of the cordilleran belt ends at Cape Horn yet the mountains of the Antarctic Peninsula have a very similar trend and structure. The similarity is so close that most geologists agree they were once united. Today they are linked across the South Atlantic by an arcuate chain of volcanically active islands that build the Scotia Arc. To this day geologists are unsure of exactly how the arc was deformed into such an enormous loop, but we can be sure that it has something to do with the tremendous forces unleashed when Gondwanaland broke up.

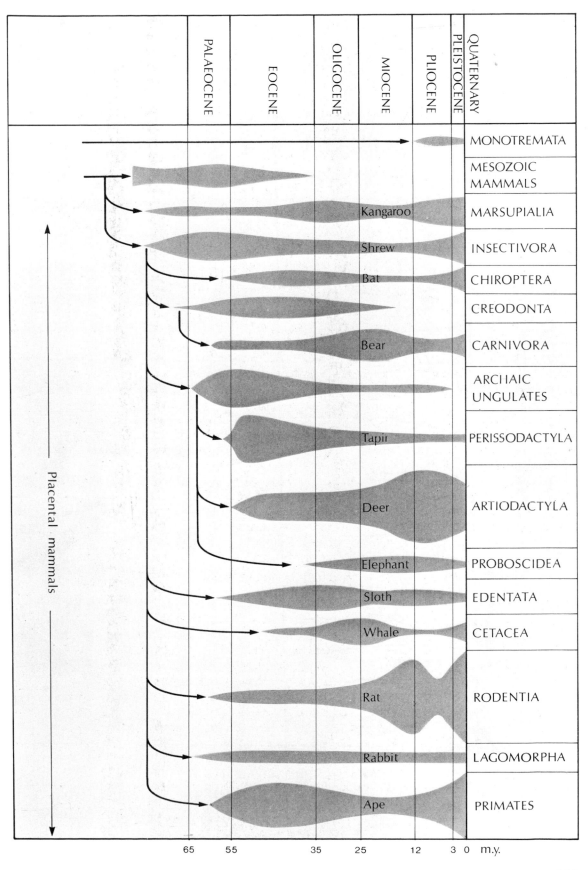

24

THE PLEISTOCENE EPOCH

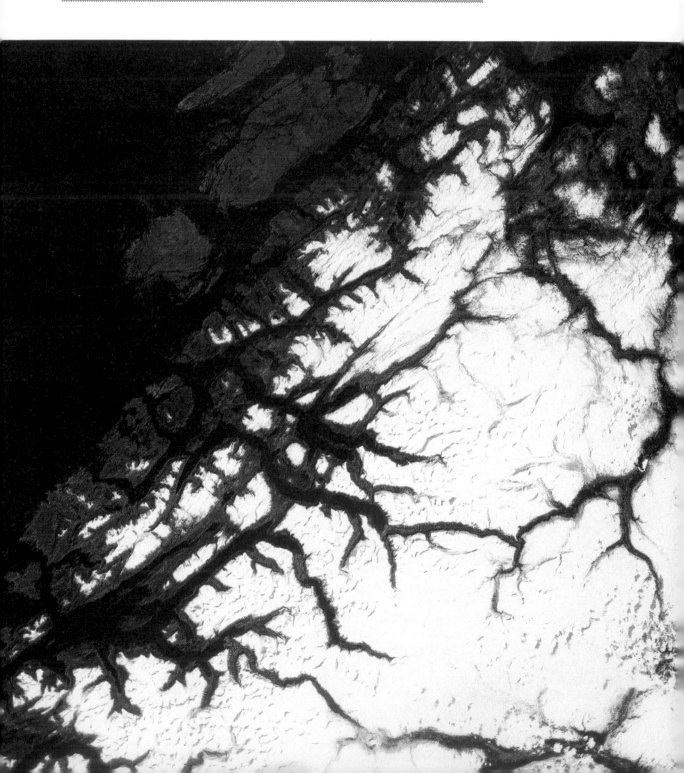

OPPOSITE Part of the glacier-carved coast and snowcap of Norway, south of the city of Tromsø (top margin of picture). All of Scandinavia was, of course, covered by the Pleistocene ice sheets and owes its intricately carved physiography to ice erosion. The mountains of the Kjølen range visible in this image rise to about 2400 m and are part of the great Caledonian chain (Lower Palaeozoic rocks).

RIGHT Pillars of boulder clay eroded from Pleistocene cliffs. Each pillar is capped by a layer of resilient material. West Wales.

The Onset of an Ice Age

We are used to regarding the Earth as a world with extremes of climate. Certainly an Eskimo would be very unhappy in Central Africa, and a Bushman would be equally unsuited to Greenland. But in fact the full range of temperatures is not nearly so great as might be imagined; a few degrees change in the global temperature can make a surprisingly large difference to the prevailing climate. In particular, it would lead to a different distribution of land and sea. If the Antarctic ice sheet melted, the mean sea level would rise by over 60 m. Such a change would require a global warming of only 5°C. At the moment, temperatures are well suited to advanced life over much of the globe, and of the large continents only Antarctica has no indigenous human inhabitants – although penguins manage very well there. However, at intervals during its history the Earth has been subject to cold spells, or glaciations. The last one ended a mere 10,000 years ago: or, in historical dating, 8000 BC, which is very recent indeed, even in terms of the era of modern Man.

As early as 1795 James Hutton, one of the founders of modern geology, suggested that the Alps had once been covered with much larger masses of ice. The idea of a former colder period became more and more widely accepted, and was finally demonstrated by the nineteenth-century Swiss geologist Louis Agassiz, who gave convincing reasons for believing that the ice sheet had covered much of northern Europe, including the whole of Scotland and part of England. Later investigations have shown that the Pleistocene glaciations were not the only ones. There have been at least six others, dating back to more than 2300 million years ago and long before we come to the start of the fossil record.

It is important to remember that a glaciation does not necessarily mean a long period of continuous cold. The last glacial period spread over at least 80,000 years, but was made up of four separate cold waves interspersed with warmer interglacials. Even now we could well be in an interglacial, in which case the ice age is not over yet.

Quite apart from this, there are more recent minor variations. Europe during the heyday of the Roman Empire was appreciably warmer than it is now. There was a colder spell between AD 400 and 800; then another warm period, which ended about 1200, during which southern Greenland was colonized by the Vikings. It was succeeded by what is often called the Little Ice Age, when climates in northern Europe were severe and the Norse settlements in Greenland were wiped out. Then, after 1780, the temperature rose again. Since 1940 there has been some deterioration, although it is not likely that this will be more than temporary.

For obvious reasons we know much less about past climates in the southern hemisphere than in the northern, but there is every reason to believe that when an ice age occurs, it affects the whole of the Earth rather than one hemisphere only. This is important when we come to consider the reasons for these cold spells.

Character of the Glaciations

It would be quite wrong to suppose that during a glacial period the whole of the Earth's surface is coated with an icy mantle. This has certainly never happened. Although the disposition of continents and oceans was quite different early in Earth history, the evidence shows that the equatorial regions of the Earth have always been ice-free even during glacial periods.

Modern Antarctica is covered with ice – and bear in mind that it is larger than the whole of Europe. Greenland, too, is mainly ice-covered. These, however, are the only major continental accumulations of ice at present. The Arctic Ocean is largely ice-covered, though this is only a relatively thin cover of drifting pack-ice. Note that there are extensive land areas in the Arctic not covered by ice, mainly because of low precipitation in these polar deserts. During greater advances of the ice, however, the polar sheets spread much further toward the Equator, so that low-lying areas in middle latitudes were also covered. The drop in world temperature produced permafrost – that is to say, permanently frozen ground – even in temperate zones, and the seas would have contained masses of floating ice in regions where currently this would seem ridiculous.

Naturally, all this has effects upon the circulation of the atmosphere, and there are also short-term changes in sea level as thinner ice sheets periodically melt and then re-form. Circulation of the surface waters of the oceans is disturbed.

But perhaps the main feature is the growth of continental ice caps. In the present interglacial period, the winter growth of an ice cap is balanced out by the summer shrinkage (as is also the case with the polar caps on Mars), but at the onset of a glacial period – the last one of which ended about 10,000 years ago – the winter growth is not reduced during the following cool summer, so that the cap grows very quickly. This in turn lowers the level of the sea, and at the height of the last glaciation, around 18,000 years ago, the mean sea level was at least 80 m below its present value. The average ocean temperature was only 2.3°C below that of today.

The steady growth of ice sheets at the start of a glaciation is not necessarily confined to the poles, although naturally the effects are greatest there. Polar regions tend to be very dry as well as very cold. Even today there are areas of Siberia, Alaska and the northern part of Greenland that are not ice-covered, whereas the more mountainous regions further south are better suited to the development of glaciers.

By mapping the deposits of ice sheets, it has been found that the glacial successions contain not only 'drift' (glacial till and outwash deposits, such as rock, sand and clay) but also soils. These have yielded plant remains, some of which palaeobotanists believe grew in relatively warm climates. This led to the inevitable conclusion that the continental ice sheets alternately advanced and retreated. During the periods of ice retreat, warm climatic conditions slowly returned to the formerly glaciated regions. Such periods are termed interglacials and there were four main ones during the Pleistocene 'Ice Age'. Some of these were of greater length than the intervening 'glacials'.

Dating the Ice Ages

The first known glaciation, called the Huronian, left deposits that date from 2700 to 1800 million years ago. The main geological evidence comes from the Lake Huron region of Canada, where three separate horizons of glacial deposits are separated by non-glacial sediments. These indicate that the glaciation was interrupted by warmer periods. From the end of the Huronian glaciations there is no further evidence for such events until 950 million years ago. Then comes evidence for further glaciations, each lasting for about 100 million years; these were the Gnejsö (940 million years ago), the Sturtian (770 million years ago) and the Varangian (615 million years ago). Our knowledge of these is extremely fragmentary.

The above four glaciations all took place well back in Precambrian time and are dependent upon radiometric dating for their correlation with the geological time scale. Later glaciations are easier to date because fossils can be used. The first Phanerozoic glaciation can be confidently dated to the Late Ordovician period on the basis of deposits in North Africa which are 480 million years old. Next came the best known of all pre-Quaternary ice ages, the Permo-Carboniferous. It lasted from 330 to 250 million years ago, probably with interglacials, and affected the whole of Pangaea, which lay partly in high southern latitudes throughout this time. There is no record of northern hemisphere glaciation at this period because no land existed there.

Subsequently global temperatures rose again, remaining very uniform and equable, and there were no more major glaciations until the

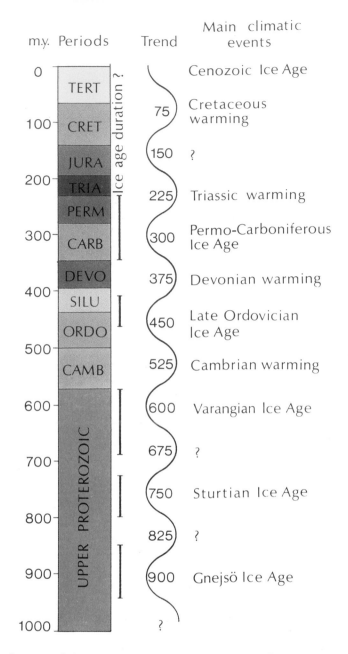

m.y.	Periods	Trend	Main climatic events

Diagram to show spacing of major glacial periods. There appears to be a 150-million-year spacing, although the hypothetical Jurassic Ice Age apparently did not occur.

establishment of the present Antarctic ice cap in the Miocene epoch. The ice cap of the northern hemisphere did not develop until later, in the Pliocene. The commencement of the Pleistocene epoch is dated at 2 million years and is entirely characterized by alternating glacial/interglacial conditions. Deep-sea cores that have been examined by oxygen isotope methods reveal there have been at least 21 glacial phases during Pleistocene times. The last major advance of the ice occurred about 16,000 years ago, but there was a brief re-advance – known as the Loch Lomond re-advance – about 11,500 years ago. This occurred very suddenly and was of but brief duration. It was primarily due to the isostatic rebound experienced by the Earth's crust after the melting of the ice sheets during the glaciation of 16,000 years ago.

Finally, one important dating method – that known as the radiocarbon method – is of particular value for dating materials which are less than 70,000 years old. This was devised in 1947 by W. F. Libby. Its use in archaeology is well-known, but it is also invaluable to geologists

Huge glacial delta fan, Syracuse, United States, showing that glaciers deposit material as well as scour the landscape.

concerned with dating very young deposits from most of the recent Pleistocene ice ages. Older isotopic methods are inaccurate for such recent ages.

The basis of the decay process is the effect on nitrogen atoms of neutrons produced in the upper atmosphere by cosmic rays. This process gives rise to the radioactive isotope of carbon: C^{14}. This isotope, together with the non-radioactive ones, C^{13} and C^{12}, is converted to carbon dioxide and is assimilated by plants during the process of photosynthesis. At death, a plant will have a C^{14} to C^{12} ratio which is roughly the same as that of the atmosphere, but the plant will then cease to acquire any more carbon. The radioactive carbon within the plant then decays. Measurement of the remaining carbon in plant material from Late Pleistocene deposits allows calculation of their age. The half-life of carbon is 5730 years.

What Causes an Ice Age?

The fact that the Earth has experienced glaciations now and then throughout its long history has been known for more than a century and a half, but we are still uncertain about the reasons for them. Many theories have been proposed, some of them decidedly far-fetched. No doubt the distribution of continents and oceans plays some part, and there are even suggestions that there may be connections with magnetic reversals of polarity, but the evidence is poor.

It seems that we might most profitably seek an explanation which is essentially astronomical, and it is natural to turn first to possible changes in the output of the Sun as a contributor to the surface temperatures on Earth.

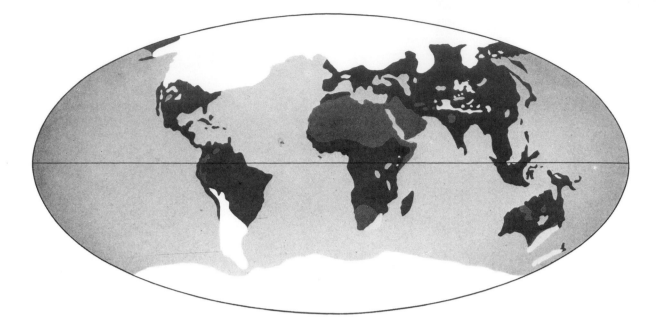

On the whole the Sun is a steady, well-behaved star. It is not always equally active; there are maxima every 11 years or so, when the bright surface shows many groups of the dark patches known (rather mis-leadingly) as sunspots, and the 'solar wind' is at its strongest. At minimum activity there may be no spots for several consecutive days or even weeks. But there seems to be no real connection between the solar cycle and the weather, and we cannot even be sure that the cycle is permanent. Indeed, between AD 1645 and 1715 the admittedly incomplete records indicate that there were almost no spots at all.

This may or may not be significant – but it is worth noting that the Little Ice Age, the most important of the medium-scale fluctuations during the Pleistocene, occurred during this period, although it began earlier (AD 1500) and finished later (AD 1900). Meanwhile, have we any evidence about more marked and long-term changes in the Sun?

Unfortunately, we have not. If the Sun's output were even slightly reduced the effects on the Earth would be catastrophic, and would certainly lead to an ice age; but we have absolutely no clue as to the state of the Sun during the Earth's glacial periods. Another untestable possibility is the suggestion that during ice ages the solar system was passing through a region of the Galaxy in which the interstellar medium was denser than usual, thereby blocking out some of the Sun's radiation.

An entirely different theory was proposed by the great astronomer John Herschel in 1830, not long after the existence of past ice ages had been demonstrated. The theory was refined and improved by the Yugoslav scientist M. Milankovich, and is usually known as the Milankovich theory. Over long periods, there are changes in both the Earth's orbit and its axial inclination. The period of precession is about 26,000 years; the period of full range of the axial inclination (known technically as the obliquity of the ecliptic) is just over 40,000 years, and there is a regular change in the eccentricity of the Earth's orbit over 90,000 to 100,000 years. All these produce slight changes in climate, and Milankovich suggested that when all three are in phase the result is a drop in the Earth's overall temperature: and, therefore, an 'ice age'. Recently (1976) this idea has received some support from analysis of deep-sea cores, which appear to indicate climatic periodicities.

25

EARLY MAN

Despite all the arguments that have raged over the years, we are still uncertain about the origin of Mankind. There are also some popular misconceptions. Many people genuinely believe that the great naturalist Charles Darwin suggested that men are descended from monkeys. Of course this is sheer nonsense; Darwin would have been the last to suggest anything of the sort. What he did say, with perfect accuracy, is that men, monkeys and apes have common ancestry. If we could trace our descent sufficiently far back, we would come to the small, largely tree-living primates of more than 65,000,000 years ago.

Moreover, even today there are some opponents of evolution. In Britain, the Evolution Protest Movement is regarded with amused tolerance, but in Arkansas and other parts of the United States there is a move to restore the teaching of creationism: the idea that Man appeared quite suddenly, in his modern form, by divine agency. The reasons for this belief are religious (or, rather, pseudo-religious) but the movement does show that still today scientists have to battle against prejudice as well as ignorance.

We must also try to decide exactly what we mean by 'man'. The search for the so-called Missing Link, a creature that is half-human and half-ape, still goes on, but fossils predating the last Ice Age are very rare and fragmentary, and the evidence is inconclusive. (There was, of course, the famous Piltdown Man, which was dug up in a gravel pit in Sussex and was classed as a true Missing Link inasmuch as it had a jaw like that of an ape, but a skull that was certainly human. Alas, Eoanthropus, or Dawn Man, proved to be a fake. The skull was of moderate age, while the jaw was that of a modern orang-utan, carefully treated so as to confuse the experts.)

Certainly there was fairly rapid evolution around 50,000,000 years ago. During the Oligocene period, which extended from 38,000,000 to 26,000,000 years ago, there were many species of monkeys and apes, classified as hominoids. There is no chance that Man is anything like as old as that, but it seems that the hominoids, from which we are descended, branched off from all other primates around 35,000,000 years ago. Then, perhaps as long as 10,000,000 years ago, there was another development: a new branch of hominoids, called the dryopithecines, became distinctly separate from the rest. These were the ancestors of modern gorillas and chimpanzees. The hominoids of this new branch could walk erect, had a small monkey-like brain and may even have used primitive tools, although here opinions differ. However, in terms of their jaw formation and dentition they were ape-like, or even man-like.

It is most unlikely that they could talk in the true meaning of the term (although even here we must be cautious in view of the recent experiments which indicate that dolphins have something which is remarkably akin to a rudimentary language). They were not hunters in an organized sense, although they must have killed animals for food. The glimmerings of civilization, such as the making of fire, were completely beyond them.

It is unlikely that we can put the age of the first true men (*Homo Sapiens*) back to more than three or four million years – although even this is much older than was believed until recently. Fragments of a Miocene-Pliocene fossil hominoid, *Ramepithecus*, have been found in India. But by the end of the Pliocene period, two million years ago, we have definitely come to the age of Man. This brings us on to *Australopithecus* and the startling discoveries made in the Olduvai Gorge.

From Olduvai to *Homo sapiens*

It has been said that the Olduvai Gorge, in Tanzania, is one of the most interesting places in the world. It was here, in 1924, that Raymond Dart found the first evidence for the hominid (truly man-like), *Australopithecus*, dating back to the Pliocene period, which ended two million years ago. *Australopithecus* was of a higher order than any ape. With a massive head, projecting jaw and small cranial cap it must have been a forbidding figure by our standards, but it was advanced enough to use stone implements. The larger species were of gorilla dimensions, the smaller ones chimpanzee-sized. Both species had an upright posture and were ground-dwellers. Their brain volume was, however, only about half that of modern man.

Other finds made at Olduvai include remains of *Homo erectus*. This Middle Pleistocene hominid was widespread in Europe, Africa and Asia and marks yet another major step forward in evolution. It is not certain whether *Homo erectus* was descended directly from *Australopithecus*; for instance, a skull has been discovered at Lake Rudolf, in Kenya, which is estimated to be three million years old and not similar to *Australopithecus*; so that there may well have been more than one line of hominid evolution.

At any rate, *Homo erectus* flourished in China and Europe as well as Africa, and apparently was well established during the earlier part of the Pleistocene epoch. Some specimens of *Homo erectus* have been found associated with *Australopithecus* in South Africa, and so it seems likely that the smaller-brained genus was only slowly replaced by its larger-brained descendant, the two co-existing for a long time.

Another find of special importance came from a cave near Zhoukoudian, 48 km southwest of Peking. This cave was inhabited for a very long time. Peking Man (the name became established before the transliteration Beijing for the name of the Chinese city was officially adopted) first lived there about 460,000 years ago, and did not vacate it until 230,000 years ago, when the cave became filled with rubble and sediment. Peking Man appears to belong to one of the same species (*H. erectus*) as came from Olduvai. He was a hunter and certainly he could make use of fire. Among the implements found in the cave are choppers and scrapers, made from materials such as rock crystal, flint and sandstone.

The earliest specimens of *Homo sapiens* are Swanscombe Man (from England) and Stuttgart Man (from Germany), both of which go back to between 250,000 and 200,000 years ago. The brow ridges of both specimens are like those of *Homo erectus*, but the back of the skull is rounded, and more modern in type than with Neanderthal Man – who appeared later, about 120,000 years ago, at or just before the start of the last glaciation.

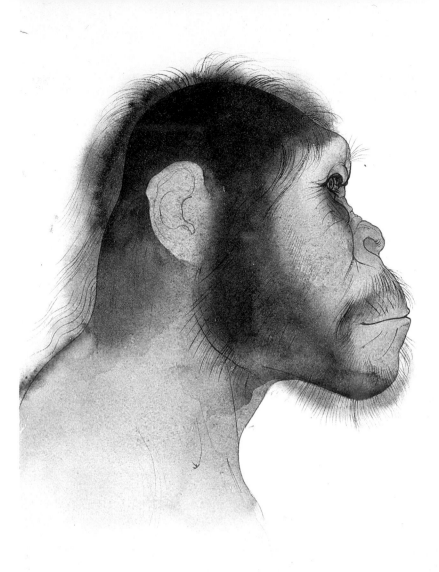

The head of *Australopithecus*.

Diorama of early man.

Neanderthal Man used to pose a problem. He was sturdy in build, with massive brow ridges; he was a hunter, and he used bone and wood to make tools. He may have clothed himself with animal skins, and even buried his dead. He is the best documented example of a heavily built race which lived about 100,000 years ago. About 35,000 years ago, when the last glaciation was at its worst, Neanderthal Man disappeared. Whether the race, which lived in both Europe and Asia, was wiped out by the arrival of the far more developed Cro-Magnon Man, we do not know. What we do now believe, however, is that both the Neanderthal and Cro-Magnon Man are too similar to modern man to merit a separate species, and that well before the end of the Pleistocene epoch (about 500,000 years ago), *Homo sapiens* had become dominant.

From Archaeology to History

The last glaciation ended 10,000 years ago: or, in historical dating 8000 BC. It is probable that true language was developed at an early stage in the existence of *Homo sapiens*. He was also a hunter on an organized scale, and as the ice began to retreat he followed the herds north.

Mammoths were one source of food – no doubt a vitally important one, as well as providing skins for clothes and even primitive housing. The fact that the mammoth became extinct can have been due to nothing but its ruthless slaughter by the hunters. Well before the end of the last glacial peak, too, we have excellent evidence that peoples who lived on the edges of glaciers in what is now Czechoslovakia and Russia built huts from poles covered with animal skins sewn together. Men knew not only how to control fire, but also how to make it. The huts were movable, but it may not be going too far to speak of 'villages', even though they were not permanent.

With the development of Neolithic cultures, which began about 9000 years ago, Man became a farmer; with the development of agriculture came the first use of domesticated animals. The rate of progress was naturally different in different parts of the world, but by 5000 years ago, humanity had spread from its places of origin across to Australia and the Americas.

Up to now we have been dealing with archaeology. Historians do not really come into the picture until we can identify the first true civilizations; the borderline is by no means clear-cut, but it may be somewhere between 6000 and 4000 BC. For instance, the Sumerians flourished by 4000 BC, and Eridu, 19 km from Ur of the Chaldees, seems to have been a permanent settlement with reed huts and mud-brick houses. Cuneiform writing had been developed before 3000 BC, and by then Egypt had been unified; according to tradition the first king of all Egypt was Menes, who had a long reign that ended when he was killed by a hippopotamus. The Old Kingdom, between about 2700 and 2150 BC, was the age of pyramid building, and nobody who has seen these immense structures can doubt that Egyptian civilization had reached a very high level indeed.

Other parts of the world also supported civilizations; in India from at least 2300 BC, in Greece somewhat earlier. By 2500 BC the Cretan civilization was well advanced and lasted for over a thousand years. Apparently it was ruined by the tremendous volcanic explosion of the island of Thira around 1500 BC. It was almost certainly this disaster that gave rise to the legend of Atlantis. Henceforth the story belongs not to the geologists, but to the historians.

26

THE EARTH NOW

Our world is an active planet, as it has continued to be ever since it was born. What more positive reminder of this fact could we have than the violent outbursts of the Mount St Helens and El Chichon volcanoes? Before this book is in print, who knows, an even more recent geological event may have hit the headlines . . .

Mount St Helens

In a report dated 1975, Mount St Helens was cited as the volcano most likely to erupt in the Cascades province of the western United States. On 18 May 1980, after nearly two months of intermittent earthquakes and steam activity, a violent nine-hour eruption sent volcanic ash into the atmosphere, devasted more than 100 km² of countryside, killed thousands of wild animals and left about 50 people dead.

St Helens is one of a number of high volcanic cones situated along the Cascades Range which extends from northern California to southern British Columbia. The chain of active or recently active centres is part of the belt of volcanicity that encircles the Pacific, and has been nicknamed the 'Ring of Fire'. Even so, activity on the scale witnessed in 1980 had not been recorded previously in the Cascades during the twentieth century, the only other eruption of note being that of Lassen Peak in 1914.

The principal destructive blast from Mount St Helens originated high on the volcano's northern flank and came with little warning. Immediately before the main eruption, a great avalanche of mixed rock, ice and mud swept down the mountain, to be followed by a violent blast that sent hot gas and rock fragments sideways into the air at hurricane velocities. The devastation associated with this event extended over 10 km from the mountain. As if this were not enough, following the avalanche, extremely hot flows of fragmented volcanic material (pyroclastic flows) and mudflows poured down the mountainside, entered stream courses and left vast deposits of sediment which completely choked the major river channels of southwest Washington state, including the Columbia River shipping channel, over 50 km distant. The cloud of volcanic ash sent out by the blast itself rose over 12 km into the atmosphere and was carried by the prevailing winds in an eastward direction, blanketing in its course vast areas of America as far east as western Montana. During the main eruptive phase, nearly 4,000,000,000 m³ of old mountain and new magmatic material was strewn over the surrounding countryside, this including a large volume of ice.

Although this devastating eruption does not necessarily mean that other Cascades volcanoes will soon erupt, it does serve as a salutary reminder that one day, others will. This is why the Mount St Helens event has become perhaps the most intensively studied and documented geological event of this decade. Only by learning all we can

The eruption of Mount
St Helens, 18 May 1980:
OPPOSITE ABOVE 8.32 am.
OPPOSITE LEFT 8.33 am.
ABOVE 8.34.20 seconds am.

about such events, can we improve our understanding of volcanoes and become more adept at predicting when and where they are most likely to occur.

Earth's Weather

Earth's normal 'weather' occurs in the troposphere. This is the lowest atmospheric layer, and it extends upwards from the surface to between 11 and 16 km. Winds are generated due to differences in atmospheric pressure from place to place over the Earth's surface. There is a general upward movement of heated air along the Equator and also at latitude 60 degrees; where such upward air movement takes place, the atmospheric pressure is relatively low. The general wind circulation of our planet is very much a function of this pattern and of the Coriolis effect, a function of the Earth's rotation. Thus, in the northern hemisphere, the winds flowing towards the Equator blow from a northeasterly direction, while those at the same latitudes in the southern hemisphere blow from the southeast. The same kind of modification of the wind pattern affects air movement at higher latitudes, where the principal movement of air is from a west to east direction. Along the equatorial belt itself, there is a zone of very gentle winds, known to mariners as the Doldrums.

On a more localized scale, the movement of air around a centre of low pressure (a cyclone or depression) is clockwise in the southern hemisphere and anticlockwise in the northern. The reverse is true of high pressure (anticyclonic) systems.

In the southern hemisphere the wind circulation at low levels is more consistent than north of the Equator. This is largely because there

is so much ocean while, in the north, the Earth's continents have a much more concentrated development.

Before the age of space exploration, meteorology was to a large extent an uncertain science. Reliable weather prediction in essence depends upon receiving reports quickly from widely separated regions of the Earth's surface. Even today neither sea nor land-based meteorological stations are evenly distributed over the surface. The whole situation was changed when it became possible to survey the Earth from very high altitudes, particularly by orbiting weather satellites. Since the 1960s these have proliferated and now play a vital part, not only in weather prediction, but also in understanding Earth's weather systems.

Photo of Earth taken in infrared by the Geos-5 weather satellite in March 1982. This shows the various circulating cloud patterns that characterize Earth's weather. The colossal plume of El Chichon volcano (seen immediately south of the Gulf of Mexico as a bright spot) spread quickly east under the influence of the prevailing southwesterly winds.

El Chichon

Because it is very remote, the 1982 eruption of the Mexican volcano, El Chichon, never made the press headlines, yet its effects were far more devastating than those of Mount St Helens. Before the event, the 1260 m cone was unimpressive. Then, at 11.32 pm on 28 March 1982, a mild earthquake shook the surrounding area. This was immediately followed by a violent explosion that ejected a column of rock and gas

Weather systems of the Earth.

17 km up into the atmosphere. A few deaths resulted. Some local people left their homes; others remained, not wishing to leave their villages. The following Saturday, 3 April, seismographs recorded over 500 tremors which gradually built in intensity until a powerful blast shook the mountain at 7.32 pm. Then, at 5.20 the following morning, the major blast occurred. This view out the core of the volcano, rather as a cork flies out of a bottle of champagne, and sent 500 million metric tons of ash hurtling into the upper atmosphere. The blast and the mantle of hot rocks and ash that spread over the surrounding region caused not only deaths to man and beast, but widespread devastation of crops, villages and communications. The precise death toll will probably never be known.

Atmospheric scientists tell us that El Chichon's volcanic cloud is the largest observed in the northern hemisphere for 70 years; we have to go back to the 1912 eruption of the Alaskan volcano, Katmai, to equal it. Its path around the world was clearly documented by orbiting satellites which photographed its movements. Most of the dust and ash has settled to the ground, but at the time of writing (1984) there is still enough remaining high in the atmosphere to cut down on the incoming solar radiation. It will continue to do so until perhaps about

where Mesozoic and Cenozoic sediments had accumulated in basins in northern Gondwanaland before it broke up. Other large fields are found along the Gulf Coast of North America and in the Caribbean. Huge resources are also being tapped in places as widely separated as Nigeria, South America and the U.S.S.R., the last probably having the largest remaining reserves.

Newer developments are currently taking place on the continental shelves, and in rifted basins such as the North Sea, where continental margin deposits have accumulated. In the case of the North Sea, more than 4 km of Permian to Recent sediments are present. Even thicker accumulations are found in the Gulf of Mexico (8 km) and in the Niger delta (7 km). The search for oil in such environments stems from the fact that more than 90 per cent of known oil and gas reserves are associated with deep evaporite deposits. As we have noted in an earlier section, such salt deposits typically form where continents have commenced to drift apart – as occurred in the case of the young Atlantic Ocean. During this stage, shallow basins form, these being cut off from water replenishment from the adjacent oceans. As plate divergence continues, so the floors of such basins slowly subside, allowing shallow-water evaporites to be deposited and to accumulate up to many kilometres in thickness. As the ocean matures and increases in area, so the bordering marine shelves continue to subside, keeping pace with the input of younger sediment. Such passive continental margins frequently provide ideal conditions for the formation of substantial oil and gas reservoirs.

ABOVE A typical oil-rig installation, Saudi Arabia.

LEFT How a salt diapir may act as a trap for hydrocarbons.

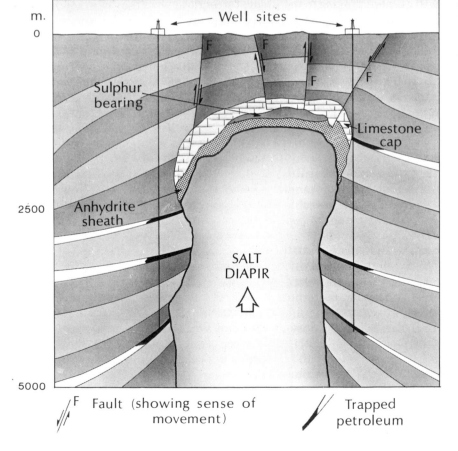

What next?

This is a book about the Earth, not about Mankind, but it is only prudent to look into the future and see whether we can foresee what lies ahead. Until geologically very recent times, Man's effect upon the environment has been negligible. Some species, such as the mammoths, have been destroyed (and let us remember that among other victims, the last of the great auks was shot less than a hundred and fifty years ago) and species of animals are disappearing at an alarming rate. Vast areas have been cultivated; this and urban expansion has meant continually increasing pressure for animals in particular, as their natural habitats have become threatened.

Today the situation is increasingly serious. A great deal is heard about pollution, which has indeed become a serious problem; it is undeniable, for instance, that the amount of carbon dioxide in the atmosphere is being steadily increased by modern industry, and if the increase continues it could do untold damage. Carbon dioxide has a 'greenhouse' effect; this is seen very clearly in the case of the planet Venus, where particularly the carbon dioxide in the atmosphere has caused the surface temperature to soar to almost 540°C. Without suggesting that our activities could produce anything so devastating as this, it could cause a rise in temperature which would change the climate very much for the worse. Even more dangerous, of course, is the possibility of a third world war, in which nuclear weapons could very easily wipe out not only Mankind, but also all other forms of life – turning the Earth into a permanently sterile, radioactive waste. Man must also beware of over-population; admittedly this is a social problem rather than a scientific one, but sooner or later Man must find an acceptable solution to it.

As to events that are wholly beyond our control, we depend entirely upon the Sun. As we have seen, the causes of Ice Ages are not known with certainty, but relatively slight fluctuations in solar output seem to provide at least one reasonable explanation, and there have also been minor changes in near-modern times which have been very marked. During the Little Ice Age, around the seventeenth century, the climate was distinctly cooler than it is now. Between 1850 and 1940 there was a warm spell, since when there has been a new deterioration. It is unlikely to last, and on balance there seems little chance of a major change in the near future. We appear currently to be in the middle of an interglacial. Such interglacials can last for well over 10,000 years, thus in the future there could well be a further glaciation. No doubt Mankind would survive, but there would presumably be a considerable reduction in world population, which would at least remove one of the potential dangers facing us!

There are other possible disasters. During reversals of the Earth's magnetic field, there may be periods when the field is virtually absent, so that harmful shortwave radiations could penetrate to ground level. There is always the risk of collision with an asteroid, which would be unlikely to destroy the world but which could alter the climate (remember the fate of the dinosaurs). Yet all in all, it seems that there is still, barring man-made disasters, a long period ahead during which the Earth will remain habitable.

Nothing can last for ever. The Sun is using up its hydrogen 'fuel', and eventually the supply of available hydrogen will be exhausted. The Sun will alter its structure; the core will shrink and heat up, while the surface layers will swell out and cool, so that the Sun becomes a red

giant star. For a period its energy output will be at least a hundred times greater than it is now. Even if the Earth survives, life here cannot. The oceans will boil and evaporate; the atmosphere will be stripped away, and the world we know will have vanished.

Luckily this will not happen for at least 4000 million years yet, and probably considerably longer. Neither is it outrageous to suggest that if humanity has survived until that remote epoch, it may have learned enough to save itself. But speculation here is not only endless, but also pointless. For the moment, we must do everything in our power to ensure that the Earth remains suited to our form of life.

BELOW The Palmerton smelter, Pennsylvania – one of the largest contributors to pollution in the USA.

BOTTOM Tailing lagoons associated with the spread of metal-polluted slurry from Benson Mines, Eastern USA. The disastrous effect upon a flourishing forest can be seen.

APPENDICES

RIGHT Eocene fish from the
Green River Shale of southwest
Wyoming.

Bibliography

We present here a selection of books into which the reader
may care to delve. Many of them are of a general nature and
each includes its own fairly extensive bibliography.

Andel, T. van. *Science at Sea*, San Francisco, Calif., 1981
    A non-technical introduction to ocean science, ranging
    in scope from continental drift to the Law of the Sea
    Conference.
Bolt, B. A. *Inside the Earth*, San Francisco, Calif., 1982
    A well-illustrated text dealing with seismology, earth-
    quakes, geophysics and plate tectonics. More technical
    sections featured in special 'boxes' within main text.
Challinor, J. *A Dictionary of Geology*, Cardiff and New York
    1978
    A much-reprinted volume of great value to geologists of
    all persuasions.
Decker, R., and B. Decker *Volcanoes*, San Francisco, Calif.,
    1981
    This book examines the geophysical background of
    volcanoes and many other related topics. An easy-to-read
    book with a wealth of information.

Dickerson, R. E. *Evolution*, San Francisco, Calif., 1978
    A good general summary of what we know about
    evolution. Amply illustrated.
Dott, R. H., and R. L. Batten *Evolution of the Earth*, New
    York 1980
    One of the best straightforward texts on historical
    geology. A lot of detail in some sections. Many maps of
    continental palaeographics.
Francis, P. *Volcanoes*, London and New York 1976
    A very readable general survey of volcanoes at a very
    acceptable price.
Gass, I. G., P. J. Smith and R. C. L. Wilson *Understanding
    the Earth*, Horsham 1974
    An Open University primer much used by universities as
    a general text for first-year students.
Gross, M. Grant *Oceanography: A View of the Earth*,
    Eaglewood Cliffs, N.J., 1982
    A highly successful introductory book that covers all of
    the main aspects of oceanography, including the
    interactions between the sea, land, atmosphere and
    marine life.

Moore, P., and G. Hunt *Atlas of the Solar System*, London and Chicago 1984
Beautifully presented, comprehensive volume, up-to-date and eminently readable. A good 'dip-in' book.

Murray, B., M. Malin and R. Greeley *Earthlike Planets*, San Francisco, Calif., 1981
Fine account of the geological features of the terrestrial planets by three of the foremost American planetary scientists.

Press, F., and R. Siever. *Earth*, San Francisco, Calif., 1982
An excellent formal text about the Earth. Much-used by universities as a first-year undergraduate reader. Good value.

Smith, D. G. (ed) *Cambridge Encyclopaedia of Earth Sciences*, Cambridge 1981; New York 1982
A technical but very wide-reaching survey of the earth sciences by a number of scientists with personal expertise in their fields.

Stokes, W. Lee *Essentials of Earth History*, Eaglewood Cliffs, N.J., 1982
A well-illustrated, rather formal but very informative book about historical geology.

Tarbuck, E. J., and F. K. Lutgens *The Earth*, Columbus, Ohio, 1983
A new and very lavishly illustrated, non-technical book about the Earth.

Scientific American books and readers

These up-to-date volumes provide excellent surveys of modern research and views about selected geological topics, most of which have been published in *Scientific American* magazine and begin with introductory remarks by a leading scientist in the particular field. They are profusely illustrated in colour and are all published by W. H. Freeman & Co., San Francisco.

Atmospheric Phenomena, 1980
Covers many topics not usually discussed in texts on earth science and meteorology.

Continents Adrift and Continents Aground, 1976
Modern compilation of material on this important topic.

The Dynamic Earth, 1984
A wide-ranging and up-to-date survey of our planet, including the interior, the crust, the oceans and the atmosphere.

Earthquakes and Volcanoes, 1980
An early illustrated compendium on these dynamic topics.

The Fossil Record and Evolution, 1982
This collection illustrates the broad range of questions and issues currently confronting palaeontologists and biologists. Many illustrations.

Life in the Sea, 1982
A collection of articles about marine organisms and the oceans in which they live.

The Planets, 1983
Ten articles outlining the current state of our knowledge of the solar system.

Volcanoes and the Earth's Interior 1982
A fully illustrated collection of articles about volcanoes and how they work. Easy non-technical style.

British Museum (Natural History): 203 (2)
Dr Peter Cattermole: 23, 24, 26, 39 (below), 69, 70, 74, 78, 81, 86 (2), 103 (below left), 107 (right), 109 (2), 111 (3), 116, 130 (2), 143 (bottom right), 144, 147 (below left), 195
Colchester Borough Council (Town Hall): 48
Bruce Coleman Ltd (Patrick Baker): 96
C. Crosthwaite: 125 (top)
C. D. Curtis: 107 (left)
Digicolor image by Daedalus Enterprises, Inc., Courtesy National Science Foundation: 67 (above)
Paul Doherty: 8, 9, 10, 14, 17, 18, 19, 21, 22, 27, 30 (above), 31, 33, 37, 41, 42, 45, 47, 50 (right), 51, (2), 52 (2), 53, 54 (2), 55, 57, (2), 58, 59 (2), 61, 62 (right), 71, 76, 77, 79 (2), 80, 83, 84, 85, 87, 88, 90, 91, 92, 94, 95 (above), 97, 100, 101, 102, 103 (above and centre), 106, 108, 110, 115, 118, 119 (after Seyfert and Serkin), 121, 126, 128 (2), 133 (after P. A. Rona, 1973), 134, 137, 138 (after B. John), 142, 145 (after J. G. Ramsay), 146, 147 (top) (after Gausser), 149, 151, 153 (after Hsu), 155 (after Molner and Tapponier), 159, 161, 165 (centre), 166, 169 (left), 172, 176, 177, 181 (top), 182-83, 184, 185, 187 (2), 190, 191, 193, 198, 200, 212, 218-19
Stan Duke: 157, 188
Earth Images (Gary Rosenquist): 206 (2), 207
General Electric Company, Space Systems Division: 189 (2)
Geological Survey and Museum, London: 139
Geological Survey of Western Australia: 95 (below)
Geology Department, University of Sheffield: 114, 127, 140 (2), 141 (2), 148, 150 (3)
Michael Holford: 50 (left)
Colin Hughes: 154, 212
P. R. Ineson 199, 214 (2)
Institute of Oceanographic Sciences: 30 (below)
Jet Propulsion Laboratory, Pasadena, Calif.: 64 (below), 66, 68 (above)
Lamont-Doherty Geological Observatory of Columbia University, New York: 68 (below), 174-75
Mark Hurd Aerial Surveys, home office: Minneapolis, Minnesota. Copyright © 1980 W. H. Freeman and Company: 34
Meteosat programme: 209
Patrick Moore: 38 (above), 39 (above left and right), 123
National Maritime Museum: 50 (left)
National Oceanic and Atmospheric Administration, Environmental Data and Information Center, National Climatic Center, Satellite Data Services Division D56, World Weather Building, Washington, D.C.: 208, 210 (2)
NERC Experimental Cartography Unit, Swindon, England: 179
B. Pigott: 28, 29 (centre and right), 46 (3), 82, 125 (centre)
Colin Reid: 124 (2), 143 (top, centre left, centre right, bottom left)
Technology Applications Center, University of New Mexico, Albuquerque, New Mexico: 62 (left), 63, 64 (above)
US Geological Survey: 165 (bottom left)
United States Department of the Interior, Geological Survey, Eros Data Center, Sioux Falls, South Dakota: 65, 67 (below), 122, 167, 169, 194
Vautier – Decool: 29 (left), 215
Vautier – de Nanxe: 2, 6, 178
Kay Whittle: 175
Woods Hole Oceanographic Institution (Dudley Foster): 181 (below)
Zefa: 156 (Dr Hans Kramarz), 168 (E. G. Carle)

Glossary

Aa: Blocky, fragmented lava.

Anorthosite: A plutonic rock (that is to say, a rock produced by igneous activity at an appreciable depth). Anorthosite is made up chiefly of calcium-rich plagioclase feldspar.

Anticline: A fold which is convex upward. On a map of an eroded anticline the oldest strata lie at the centre.

Arroyo: A steep-sided, flat-bottomed gully in a dry region with a stream that flows only at certain seasons.

Basalt: A fine-grained, dark igneous rock made up principally of pyroxene and plagioclase feldspar.

Basic rock: A rock containing minerals rich in iron and magnesium, but no quartz.

Batholith: A large, irregular mass of igneous rock, coarse-grained, intruded into the surrounding rock.

Breccia: A sedimentary or volcanic rock made up chiefly of large angular fragments.

Butte: A steep-sided, flat-topped hill made up of flat-lying strata in a deeply dissected landscape.

Caldera: A large circular volcanic depression, formed by explosion, (maybe modified by erosion), or by the collapse of the surface due to material being withdrawn from below. Many volcanoes have summit calderas.

Canyon: A deep valley with steep sides, formed by a river.

Conglomerate: A sedimentary rock, containing rounded boulders and pebbles.

Continental drift: The movements of continents relative to each other.

Continental shelf: The gently sloping, submerged edge of a continent.

Convection: The transfer of heat in a fluid material. Hot material rises because of its lower density, while colder material sinks. Convection currents occur in the Earth's mantle.

Core: The central part of the Earth, below 2900 km. It is made up largely of iron and nickel. The inner part is probably solid, the outer part certainly liquid.

Craton: A part of a continent which has not been subject to any major deformation for a long time; it may have a thin covering of younger rocks.

Crust: The outermost layer of the lithosphere. Continental crust is largely granitic, the ocean crust basaltic.

Crystal Material in which the atoms (or molecules) are regularly arranged to form a regular network.

Curie point: The temperature above which a particular magnetic mineral loses its magnetization.

Delta: Fan-shaped accumulation of sediment deposited in a sea or lake at the mouth of a river.

Diatom: A single-celled plant with a siliceous framework, found in the surface waters of seas and lakes.

Differentiation: The process of separation of heavier from lighter materials in a planetary body, so that it eventually becomes layered.

Dip: The maximum angle at which the inclination of a stratum differs from the horizontal.

Dolomite: A carbonate mineral made up of calcium, magnesium, carbon and oxygen (e.g. $CaCo_3$, calcite).

Dune: Mound of wind-blown sand formed by wind action.

Dyke (or Dike): A sheet-like body of intrusive igneous rock whose boundaries cut across bedding planes or other structure planes in the host rock.

Eclogite: A metamorphic rock containing pyroxene and garnet, formed only at very high pressure.

Epicentre: The point on the Earth's surface directly above the focus of an earthquake.

Era: A division of geological time. The eras of the Phanerozoic Eon are the Palaeozoic, Mesozoic and Cenozoic.

Erosion: The process in which soil and rock are loosened, broken down and worn away.

Extrusive rock: Volcanic rock, i.e. igneous rock that has emerged at the Earth's surface. Includes lava and pyroclastic rocks.

Fault: A fracture in the Earth's crust, across which there has been relative displacement.

Fault plane: The plane of a fault.

Feldspar: Silicate minerals usually containing sodium, calcium or potassium, and whose crystalline structure is a 3-D framework. The most important group of silicates.

Fiord (or Fjord): Steep-sided coastal inlet formed by glacial deepening of a pre-existing river valley.

Focus: The point at which an earthquake originates, well below the Earth's surface.

Fold: A flexure in rocks, identified and described by the change in dip across it.

Fossil: An impression, cast, outline, track, or part of the body of an animal or plant, preserved in rock.

Fumarole: A small volcanic vent sending out gases and hot water, but no lava.

Gabbro: A black, coarse-grained, intrusive igneous rock, containing feldspars and pyroxene.

Geomorphology: The study and interpretation of landforms.

Geosyncline: A major accumulation of sediments with or without volcanic rocks. A largely obsolete term.

Geyser: A hot spring which violently expels hot water and steam. The heat is normally due to contact between groundwater and magma.

Glacier: A river of ice which flows slowly downhill under its own weight.

Gneiss: A coarse-grained metamorphic rock with a characteristic banded texture.

Graben: A downthrown block between two normal faults; a rift valley.

Gully: A small steep-sided valley, or a channel produced by erosion.

Half-life: The time required for half of a given quantity of a radioactive isotope to decay. It is not affected by external forces or conditions, and thus forms the basis of radioactive dating.

Horst: An elevated block bounded by parallel normal faults, which may form a topographic ridge or plateau.

Igneous rock: Rock formed by the solidification of magma.

Index fossil: A fossil of particular value in stratigraphic time correlation which can be dated with exceptional precision.

Intrusion: A body of igneous rock formed from magma which has forced its way into fractures in pre-existing rocks.

Isostasy: The equilibrium between crustal blocks of differing thickness and the underlying denser mantle.

Isotope: One of several forms of one particular element which differ by virtue of differing numbers of neutrons. Thus carbon has three main isotopes, C^{12} (the most abundant), C^{13} and the radio-active C^{14}.

Laccolith: A dome-shaped igneous body intruded into formerly horizontal strata.

Lapilli: Fragments of volcanic rock produced when magma is thrown into the air by expanding gases.

Lava: Molten rock (magma) extruded at the Earth's surface, where it may spread out or flow before solidifying as a *lava flow*.

Limestone: A sedimentary rock made up chiefly of calcium carbonate.

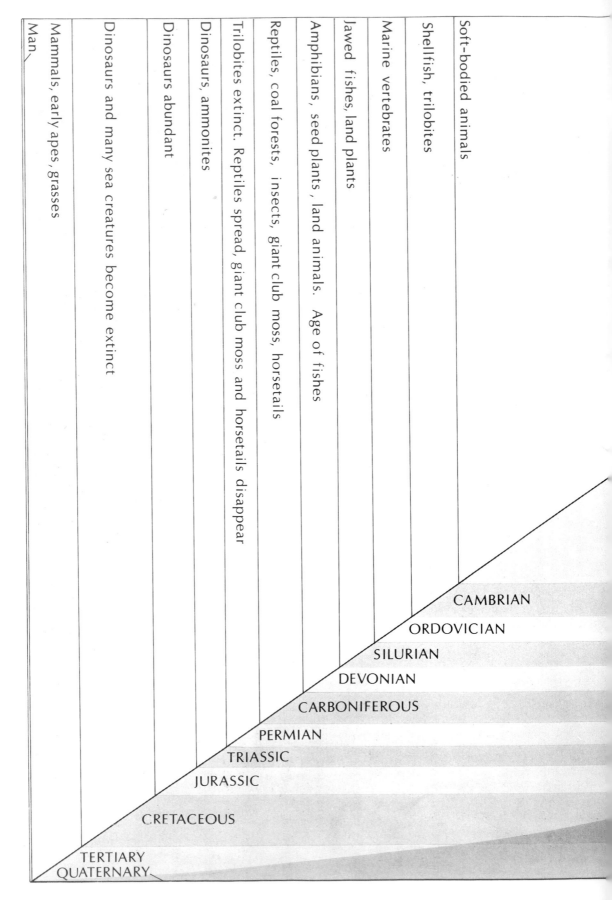

Soft-bodied animals

Shellfish, trilobites

Marine vertebrates

Jawed fishes, land plants

Amphibians, seed plants, land animals. Age of fishes

Reptiles, coal forests, insects, giant club moss, horsetails

Trilobites extinct. Reptiles spread, giant club moss and horsetails disappear

Dinosaurs, ammonites

Dinosaurs abundant

Dinosaurs and many sea creatures become extinct

Mammals, early apes, grasses
Man

CAMBRIAN

ORDOVICIAN

SILURIAN

DEVONIAN

CARBONIFEROUS

PERMIAN

TRIASSIC

JURASSIC

CRETACEOUS

TERTIARY
QUATERNARY

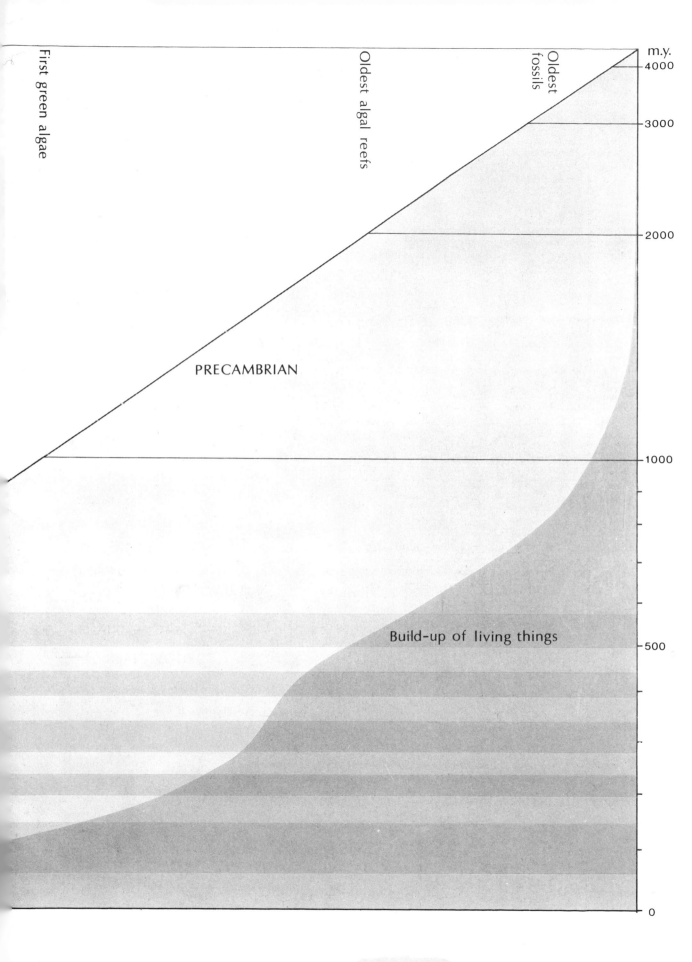

First green algae

Oldest algal reefs

Oldest fossils

PRECAMBRIAN

Build-up of living things

m.y.
4000
3000
2000
1000
500
0

Lithosphere: The outer rigid shell of the Earth. It includes the continents and their plates as well as the oceans. It includes the whole of the crust and the uppermost solid part of the mantle.

Magma: Molten rock which forms igneous rocks when cooled. Magma reaching the Earth's surface as lava is usually a mixture of fluid and crystalline material.

Magnetic reversal: A change in the polarity of the Earth's magnetic field.

Mantle: That part of the Earth between the crust and the core; it extends from the base of the crust (10–80 km) to about 2900 km.

Matrix: Fine-grained material of an igneous or sedimentary rock in which are embedded coarser fragments or cystals.

Mesa: A flat-topped, steep-sided upland; erosion of a mesa results in a butte.

Metamorphic rock: Rock formed by alteration of a pre-existing rock under conditions of increased temperature and/or pressure but without any melting taking place.

Mohorovičić discontinuity (the 'Moho'): The seismic boundary between the crust and the mantle of the Earth. It lies between 5 and 45 km below the surface, and represents a significant charge in P-wave velocity.

Moraine: A sedimentary deposit left at the edge of an ice sheet or by the retreat of a glacier.

Normal fault: A fault in which the fault plane dips towards the downthrown side of the fault.

Ore: A natural deposit from which one or more metals can be extruded economically.

Orogeny: The sum of tectonic processes in which large belts of the Earth's crust are folded, thrust-faulted, metamorphosed and intruded with igneous rocks.

Outcrop: The area over which a rock body reaches the Earth's surface.

Pahoehoe: Basaltic lava with a ropy surface.

Palaeogeography: Reconstruction of the Earth's surface (e.g. the distribution of land and sea) in past ages.

Palaeomagnetism: The study and interpretation of ancient magnetic fields recorded in rocks, and hence of changes in the Earth's magnetic field in past ages.

Palaeontology: The study of fossils and hence of ancient life.

Peléan eruption: A volcanic eruption with great explosions and the emission of glowing clouds (nuées ardentes) of gas and suspended volcanic fragments.

Peridotite: A coarse-grained and very dense igneous rock rich in iron and magnesium, made up chiefly of olivine (a silicate mineral). Thought to be an important constituent of the upper mantle.

Permafrost: Permanently frozen ground or bedrock, of which the surface layer may thaw in summer.

Reverse fault: A fault in which the fault plane dips towards the upthrown side of the fault.

Sandstone: Sedimentary rock made up of grains from 1/16 to 2 mm in diameter, bound together with a cement. The most common grain types are quartz, feldspar and fragments of rock.

Schist: Coarse-grained metamorphic rock in which the minerals are platy (such as micas) and crystallized parallel to each other.

Scoriae: Cinder-like fragments of pyroclastic rock.

Sediments: Unconsolidated particles of minerals and rock fragments deposited at the Earth's surface by wind, water or ice, or by chemical or biological processes.

Sedimentary rock: Rock formed by the accumulation and cementation of sediment.

Seismology: The study of earthquakes.

Shale: A fine-grained sedimentary rock made up of clay and silt, and which is laminated (finely layered).

Shield: A large region of stable, ancient rocks, outcropping at Earth's surface (e.g. Canadian Shield).

Slip (fault): The movement of one face of a fault relative to the other.

Stoping: A theoretical process in which an intruding mass of magma makes room for itself by breaking off fragments from the country rock through which it is ascending, and assimilating them as they sink.

Stratigraphy: The science of stratified rocks and their interpretation in terms of geological history.

Stratum: A layer (or bed) of sedimentary rock.

Strike: The angle between true north and the horizontal line contained in any planar feature (such as a stratum).

Strike-slip fault: A fault along which the displacement is predominantly horizontal (i.e. parallel to the strike).

Stromatolite: A fossil form of dome-like structure generally made up chiefly of calcium carbonate arranged in parallel laminae, probably precipitated or accumulated by blue-green algae. Stromatolites are particularly important in Precambrian strata and have a range well back into the Archean. They are alive at the present day.

Sublimation: The change of state of a substance direct from solid to gas.

Syncline: A flexure in the form of a trough. In an eroded syncline, or on a geological map of a syncline, the youngest strata are in the centre (cf. anticline).

Tectonics: The study of large-scale structural features of the Earth's crust and their interpretation in terms of Earth movements.

Thrust fault: A reverse fault in which the dip of the fault plane is shallow.

Till: An unconsolidated sediment made up of fragments of very variable size, deposited by glacial action.

Transgression: A rise in sea level relative to the adjacent land, causing formerly exposed areas to be submerged and subject to sedimentation.

Trench: A long, narrow, deep trough in the ocean floor. Characteristically found on the oceanward side of island arcs, and marking the site of convergent plate boundaries.

Trilobites: Extinct crustacean arthropods. They first appeared at the beginning of the Cambrian Period, and then died out in the Permian.

Tsunami: A 'tidal' wave, due not to tides but to submarine earthquakes.

Tuff: Consolidated rock composed of pyroclastic fragments and fine ash.

Unconformity: A surface which separates strata of significantly different age and thus represents an interval of non-deposition. An angular unconformity indicates a period of folding and erosion between time of deposition of under- and overlying strata.

Vein: A deposit of minerals within a rock fracture.

Vesicle: A cavity of an igneous rock formerly occupied by a bubble of gas formed by decompression of the magma as it rose towards Earth's surface.

Volatiles: Those components of magma which are readily lost as gases when the pressure on the magma is reduced, as when it rises through the crust.

Volcano: Any vent or fissure in the earth's crust through which magma, gases and pyroclastic debris are erupted.

Xenolith: A fragment of a pre-existing rock found inside an intrusive igneous body. Such fragments may be incorporated as magma passes through crustal fractures.

Index